# Data Product Management in the AI Age

## Design and Manage Your Data Strategy to Get Ahead

Jessika Milhomem

Apress®

*Data Product Management in the AI Age: Design and Manage Your Data Strategy to Get Ahead*

Jessika Milhomem
Sao Paulo, Brazil

ISBN-13 (pbk): 979-8-8688-1314-6          ISBN-13 (electronic): 979-8-8688-1315-3
https://doi.org/10.1007/979-8-8688-1315-3

Managing Director, Apress Media LLC: Welmoed Spahr
Acquisitions Editor: Shaul Elson
Development Editor: Laura Berendson
Coordinating Editor: Gryffin Winkler
Copy Editor: Kim Burton

Cover Image by Racool_studio on Freepik (www.freepik.com)

Distributed to the book trade worldwide by Springer Science+Business Media New York, 1 New York Plaza, New York, NY 10004. Phone 1-800-SPRINGER, fax (201) 348-4505, e-mail orders-ny@springer-sbm.com, or visit www.springeronline.com. Apress Media, LLC is a Delaware LLC and the sole member (owner) is Springer Science + Business Media Finance Inc (SSBM Finance Inc). SSBM Finance Inc is a **Delaware** corporation.

For information on translations, please e-mail booktranslations@springernature.com; for reprint, paperback, or audio rights, please e-mail bookpermissions@springernature.com.

Apress titles may be purchased in bulk for academic, corporate, or promotional use. eBook versions and licenses are also available for most titles. For more information, reference our Print and eBook Bulk Sales web page at http://www.apress.com/bulk-sales.

Any source code or other supplementary material referenced by the author in this book can be found here: https://www.apress.com/gp/services/source-code.

If disposing of this product, please recycle the paper

# Table of Contents

# About the Author

**Jessika Milhomem** is a Data & AI leader with nearly two decades of experience helping companies turn data challenges into strategic growth. She specialized in bridging technical depth with business impact, performing her work by mixing her hands-on technical skills, cross-domain data strategy and product mindset. She has been leading high-performance teams in national and international projects across a wide range of industries, such as digital banking, digital fashion, telecommunication, education, automotive and consumer goods.

Her passion lies in integrating the triad of business contexts and needs, the power of data, and the team effort dynamic to deliver impactful results. That's why she works close to her teams and leads by empowering them to consume and produce data products and data as a product with strong strategy as a foundation to strengthen business opportunities and decisions in creating innovative customer solutions.

You can find more information about her work at `https://jessikamilhomem.com`.

# About the Technical Reviewers

 **Kent Graziano**, a.k.a. The Data Warrior, is an internationally recognized industry thought leader, award-winning speaker, author, and semi-retired Snowflake Data Cloud and Data Vault evangelist with too many decades in the industry to count! He is a Data Vault Master, Knight of the Oaktable, Oracle ACE Director – Alumni, and Taekwondo grandmaster.

Kent is a recognized expert in Snowflake Data Cloud, agile data warehousing, and agile data modeling. He has written numerous articles and blog posts, authored three Kindle books (available on Amazon.com), co-authored four other data-related books, and has done hundreds of presentations and podcasts nationally and internationally. Kent was a co-author of the first edition of *The Data Model Resource Book*, the first data vault book (with Dan Linstedt), and the technical editor for *Super Charge Your Data Warehouse* (the main technical book for DV 1.0).

 **Lucas Bittencourt Xavier** is a data architect with extensive experience designing and building scalable data solutions for startups and large enterprises. He specializes in data architecture, modeling, and automation, creating efficient systems that transform raw data into valuable insights.

Passionate about data quality and performance, Lucas has worked across the full data lifecycle, from designing infrastructure to optimizing analytics. He enjoys solving complex data challenges and continuously improving systems to support business growth.

When he's not working with data, Lucas is constantly exploring new technologies and industry trends.

# Acknowledgments

I dedicate this book to God, who always gave me physical and mental capabilities and opened all my opportunities. And to myself for believing in me and always fighting for my goals with love, obstinacy, resilience, and diligence!

I would like to wholeheartedly thank my beloved and loving husband, Ricardo, who is my partner in life from so many angles and perspectives. Of course, he read my book many times and gave me a lot of feedback! He was my first reader and editor!

I appreciate everything you do! I love you very much and the life we have been developing together during these long years. And there are so many that we'll come!

Thank you to my parents, Beth and João, who always did what they could and couldn't, not just for my existence but also for my intellectual development and values, as well as they did for my brothers! And you always did it with much love. I'm a warrior because I have wonderful teachers in my home! Thank you for the values you taught me! I love you!

Thank you to my brothers, Bruno Vinicius and Erika Araceli, for being my best friends, rooting for me, and motivating me whenever I needed it! You are very important to me, and I'm super proud about you both! I love you!

Enzo, Emma, Eric, my beloved nephews, I hope you see this book as an inspiration to dream big and work hard to achieve them! It's completely possible, although eventually hard (or super-ultra hard)! Everything we do with dedication, passion, diligence, and hard work is achievable! We need to do our part while God does them! I love you so much!

## ACKNOWLEDGMENTS

I would also like to thank my fifth-year Portuguese language teacher, Cristina Oliveira, who encouraged me to develop a greater love for reading.

Finally, I'd like to thank my editors, Shaul Elson, Gryffin Winkler, and Laura Berendson, who carefully guided me throughout this writing adventure! I do appreciate your work! You're amazing!

Also, I would like to thank my technical reviewers, Kent Graziano and Lucas Bittencourt, for all the detailed and fantastic collaborations carried out during this journey!

# Introduction

## The Motivation for This Book

You are in the all-hands meeting of the company you work for, and the C-level team is delivering a very inspirational pitch about the importance of innovating by using more and more artificial intelligence. They share some of the projects already in progress or provide a high-level overview of projects that will be implemented in the company. The company's professionals are inspired and very excited with the pitch, especially the senior leadership. After all, the company may achieve great results with all these innovations!

Now, you attend the business unit's leadership meetings to discuss the strategy planning for the following cycle. The teams have planned significant activities for their respective business settings to accomplish a common aim. Some are also considering novel approaches to applying data models and leveraging artificial intelligence for answers. Everything is mapped except the data quality initiatives you formed agreements or advised that they undertake. After drawing attention to it, the leaders state that what they have planned is sufficient for their key aims and it can accomplish the desired results alone through self-service data usage. Coincidentally, the individual contributors from their teams share the same mindset and will just need collaborative work with the data individual contributors.

At another moment, a leader of your business area asks for your help to investigate issues in the data of the teams you've just aligned with. They shared it with another business unit to enable the implementation of a new strategical business process. However, they have faced and are again impacting the business because the data results are mismatched from the expected behavior. It is unclear whether any process failed, the data is wrong due to a bug, or if there's another issue involved. Even worse, they don't know where to start the investigation.

Then, one day, a leader from another Business Unit needs to report consolidated company-level information to the C-level team or to a regulatory agency and comes to you asking for a source of true analytical data from your business domain to ensure the accuracy of the information. However, no such thing exists because each team works with their data using different approaches and rules, and none end-to-end data strategy was designed and considered in any roadmap Consequently, they either have partial information or none at all, as the rest of the areas, including you. However, even though, you and your team still need to work on that super innovative projects of AI that demands data, and are the expectation for the company's interests, that were already set up and communicated by the C-level for the company. And everything is important for the company. All the business needs are real.

Could you see yourself or someone in at least one of these scenarios? I imagine that they might not be uncommon to you. And yes, we could talk about many other "hypothetical scenarios"!

While there is no unique silver bullet to address all of them, concrete strategy and practical work designed to manage data products (from data to AI) is fundamental to ensure all the needs and expectations of the business. And it's critical that leaders have knowledge and experience to deal with it. For this, some skills and frameworks based on important fundamentals can and should be applied to address these challenges. It is the huge opportunity I saw to share the knowledge and techniques I developed throughout my professional career journey over almost two decades in different markets, also experiencing real-world cases with diverse businesses contexts, technical environments, and professionals (mid-senior managers/directors and individual contributors) from different markets.

We're in the era of artificial intelligence. The backbone of this technology is data by providing the content learned, while the algorithm is the brain that guides the muscles to understand the paths.

In other words, data is critical for any strategy nowadays. Thus, robust data product management is fundamental. If the data used is inaccurate or biased in some aspect, the performance of AI systems will be directly affected, as they may learn incorrect patterns. The potential impacts of this can be significant!

In 2021, Gartner stated: "Every year, poor data quality costs organizations an average of $12.9 million. Apart from the immediate impact on revenue, over the long term, poor quality data increases the complexity of data ecosystems and leads to poor decision making."[1]

To succeed, companies need robust data architectures that facilitate self-service analytics, enabling faster product tests and implementations in the short term. These initiatives may not necessarily require best-in-class governed and accurate data.

However, companies must also be capable of successfully scaling emergent solutions into consolidated ones within healthy platforms with foundation capabilities for the middle and long term.

Besides the obvious continuous discoveries for new business opportunities, any company has responsibilities for internal and external purposes that require curated data, fundamentally as the canonical data (a.k.a source of truth data).

These responsibilities entail the following internal and external needs.

- Supporting the design, development, delivery, and maintenance of solutions with scalability, including monitoring, policies, and AI/machine learning (ML) models

---

[1] Gartner.com (2021), *How to improve your data quality*. Available at https://www.gartner.com/smarterwithgartner/how-to-improve-your-data-quality

- Providing efficient routine (business as usual) among professionals

- Upholding governance standards for internal and external purposes

- Feeding enterprise mechanisms that deliver information to customers

- Reporting ad hoc or recurrently reports to regulatory agencies

Due to these reasons, it is critical to have responsible professionals for the data strategy of businesses who know how to design and implement data products and data as a product.

# Poor Data Quality Definition

The poor data lacks conformance of business rules, completeness of data, format compliance, consistency rules, accuracy verification, uniqueness verification, and timeliness validation (DAMA International, 2015).[2] This results in one and/or more of the following characteristics.

- Inaccurate data

- Duplicate data

- Incomplete data

- Non-compliant data

It can also be generated due to the lack of ownership and leadership of data.

---

[2] DAMA International, (2015). *Data DMBook*, Technics Publications p. 591 to 601

- Lack of awareness on the part of leadership and staff

- Lack of business governance

- Lack of leadership and management

- Difficulty in justification of improvements

- Inappropriate or ineffective instruments to measure value

# Poor Data Quality Impacts Business

The poor data quality directly impacts the business, even if it is not always easy to demonstrate or if it is explicit.

These issues can lead to risks or impacts in at least four aspects.

- **Regulatory violations and fines** (e.g., reporting bad data) result in getting fined

- **Higher operational costs** from unfit processes using inefficient and non-standardized data architectures

- **Loss of (opportunities to generate) revenue** from not understanding the market and internal opportunities to generate revenues, reduce costs and losses

- **Negative reputational impact** from things like when companies announce they miscalculated something (e.g., underpaid compensation costs by Uber to its drivers[3])

---

[3] Business Insider (2017), *Uber is paying its NYC drivers 'tens of millions' because of an accounting error that underpaid them for years.* Available at https://www.businessinsider.com/uber-paying-nyc-drivers-tens-of-millions-accounting-error-2017-5

# The Purpose of This Book

Accurate and very well-managed data is important, and it is not just for companies. It is fundamental for a lot of businesses.

It becomes even more critical now, in the initial era of AI, as data is the information that AI uses to learn and make decisions while using algorithms as rules to follow, process data, and learn from (David Pattel, 2023).[4]

With that in mind, this book aims to help you perform this important responsibility of managing data as a product by helping you to answer important questions.

- What is a data product, and why is it necessary?
- What is data as a product, and why is it necessary?
- What should I know to do this work?
- How should I design it?
- How should I get started?
- What and how should I do it?

*Data Product Management in the AI Age* is a book intended for several types of readers.

- Data executives who will lead the design and implementation of data products
- Product managers interested in becoming data product managers/owners
- Business people who will work with data teams, managing data products and their projects,

---

[4] David M Patel (2023). *Artificial Intelligence & Generative AI for Beginners: The Complete Guide.* Kindle Edition, p.23.

- Analytics engineers, data engineers, data scientists, and/or machine learning engineers who will implement data solutions

- Data-related professionals and developers who are aspiring data leadership and product management

The book does not presume a sophisticated agile and project management background. However, by its very nature, the material is grounded on its concepts.

The goal is to impart a significant understanding of practical data product management and, more importantly, give practical guidelines about how to perform it.

To address this, the book covers the following.

- **Introducing data product management fundamentals**: You'll learn about the concepts of the products, what are data products, data as a product and the origin of everything, the analytical architecture journey and its issues, and the necessary knowledge to manage it.

- **Putting data product management into practice**: You'll learn new techniques I designed, such as the Golden Data Platform, used to manage data as a product through a data product. And you'll also learn about the data product management framework and how to use the canvas I created. Finally, you'll learn how to put everything into practice, from the designing of the solution to the ongoing of it.

By the end of the book, you will be able to explain to your leaders, peers and team why data products and especially data as a product are important for the business. Not just explain it, but put it into practice and deliver value to the business in this new era!

## INTRODUCTION

Although it's not mandatory, I recommend you read the chapters sequentially for a deeper understanding of the topics approached and a gradual learning path.

# PART I

# Fundamentals for Data Product Management

# CHAPTER 1

# Introduction to Data Product Management

In this chapter, you'll learn the main concepts of data product management.

I'll explain what a product is and define data management, data governance, artificial intelligence, data products, data as a product, and how everything is related.

This chapter contains fundamental concepts that many experienced professionals still don't know and are fundamental for any seniority of professionals.

## 1.1. Definitions

Before talking about product definitions, it is fundamental to understand the definition of a problem. After all, products are created upon them. At least, they should be.

According to Collins Dictionary: "A problem is a situation that is unsatisfatory and causes difficulties for people."

Dictionary.com describes a problem as "1. any question or matter involving doubt, uncertainty, or difficulty. 2. a question proposed for solution or discussion."

© Jessika Milhomem 2025
J. Milhomem, *Data Product Management in the AI Age*,
https://doi.org/10.1007/979-8-8688-1315-3_1

As you can see, the problem definition is also an invitation to create a solution, which is the main core of the product's definition.

Now, let's use the same approach and evaluate the definition of a product through the dictionary!

Dictionary.com describes the product definition as follows: "1. a thing produced by labor. 2. a person or thing produced by or resulting from a process, as a natural, social, or historical one; result. 3. the totality of goods or services that a company makes available; output."

Scrum,[1] one of the most important agile[2] frameworks today, states, "A product is a vehicle to deliver value. It has a clear boundary, known stakeholders, and well-defined users or customers. A product could be a service, a physical product, or something more abstract."

This is the foundation concept for this book.

I enjoy the concept description the most because it represents the core of the product's expectations and what every work result should deliver: *value—* the foundation of everything I've been discussing since this book's beginning!

It means that a product is also measurable. It must enable us to measure the result of a problem solved by the product, even if it is a complex job to be done!

With that in mind and recapping the initial discussion, the main principle of a product is **to solve a problem for someone.** For this reason, it is important to be grounded in two main pillars when designing a product.

- The problem: Map the issue and its full context.

- The target audience: Map the interested public in the solution and its motivations.

---

[1] Simon Kneafsey (2014), "What is a product?" Scrum.org. Available at https://www.scrum.org/resources/blog/what-product#:~:text=%E2%80%9CA%20product%20is%20a%20vehicle,,%20or%20something%20more%20abstract.%E2%80%9D&text=This%20is%20an%20intentionally%20broad,and%20applicable%20in%20multiple%20contexts

[2] Agile is a set of methods and practices where solutions evolve through collaboration between self-organizing, cross-functional teams.

# 1.2.  Data Products

Products exist because there are opportunities. In other words, there are issues/problems to be solved. It also applies to the data world.

In 2021, Gartner, one of the biggest consulting companies around the world that evaluates technologies solutions for decision-making and strategies of companies, shared that "Poor data quality costs organizations an average of $12.9 million. Apart from the immediate impact on revenue, over the long term, poor quality data increases the complexity of data ecosystems and leads to poor decision-making."[3]

In 2024, Fivetran and Vanson Bourne performed new research[4] with more than 500 companies from the United States, Germany, France, the United Kingdom, and Ireland that are operating in the private and public sectors. The results show that although 97% of the companies are investing in artificial intelligence (AI)/machine learning (ML) models in the next two years, 81% of them trust their AI/ML outputs, admitting they have data foundation inefficiencies, which are necessary to improve their business outcomes.

The data issues come from technical and non-technical aspects.

The following are technical aspects.

- Disorganized and siloed data

- IT infrastructure outdated

- Low quality of data

- Stale data

- Access to the data

---

[3] Manasi Sakpal (2021), "How to improve your data quality." Garter.com. Available at https://www.gartner.com/smarterwithgartner/how-to-improve-your-data-quality

[4] Fivetran (2024), *Fivetran + Vanson Bourne report: AI in 2024.* Available at https://www.fivetran.com/resources/reports/fivetran-vanson-bourne-report-ai-in-2024

The following are non-technical aspects.

- Lack of buy-in and support from senior teams

- Lack of internal skills

Consequently, 40% of the companies admitted they had experienced data inaccuracies, hallucinations, and data biases in their AI outputs.

Moreover, in financial aspects, underperforming AI programs/models built on inaccurate or low-quality data cost 6% of global annual revenue, which represents approximately $336 million a year, based on data from 550 respondent organizations with more than $25 million annual revenue and an average of $5.6 billion.

Data quality becomes a key barrier to good business results. Especially as organizations reach the advanced stage, in a nutshell, for AI models to produce the best output and create impact for the business, generating the necessary value result, the data used must be of the foremost quality.

# Data Management Fundamentals

Before talking about the types of data issues, it is critical to understand some data management fundamentals.

Unfortunately, during my almost two decades of experience, I'm used to seeing lack of knowledge as a huge deficiency for mid/senior data professionals (managers and individual contributors). This is one of the reasons for this book.

These concepts are fundamental not just for operational and tactical executions of any project, initiative, or product but for strategic thinking. After all, managing such important assets as the company's data requires a lot of strategic thinking and designing. At least, it should.

# What Is Data Management?

In the 1950s, database theory and usability started with the data storage's existing solutions. The technologies kept evolving until the invention of relational databases in the 1970s, which were used as repositories until the present.

In the 1980s, databases started to be used for analytical purposes (traditionally nominated as decision support systems) as database marketing[5] designed to innovate and improve direct marketing[6] because it was previously done offline with strong manual effort.

Sequentially, the data warehouse concept emerged, engaging more business areas to leverage data organization and usability, and so on.

Due to the power of data, companies have started to recognize their data as a vital enterprise asset, enabling innovations and strategic achievements. Due to its importance, it became inefficient for the business to manage data in an ad hoc manner based on which kind of value it could derive. For this, it was necessary to change the way to manage data: with focus, intention, planning, leadership, and commitment.

According to DAMA International:[7] "Data management is the development, execution, and supervision of plans, policies, programs, and practices that deliver, control, protect, and enhance the value of data and information assets throughout their life cycles."

Thus, data management is the processes and procedures required to manage data. Its main goal is to manage data as a valuable asset.

---

[5] Database marketing is a sort of direct marketing in which a company's client databases are utilized to create customized email lists for marketing campaigns, a practice known as customer segmentation.

[6] Direct marketing relies on one-on-one communication with a target audience. It includes tools like emails, phone calls, catalog marketing, and text messages.

[7] DAMA International, (2015), *Data DMBook*, Technics Publications.

With the data management necessity emerging, the DAMA-DMBoK (DAMA International's Guide to the Data Management Body of Knowledge) framework was designed in 2009 to enable structured data management, which received updates in 2015 to englobe new needs related to big data technologies.

## Data Governance

Piethein Strengholt (2020)[8] says that data governance "consists of activities for implementing and enforcing authority and control over data management, including corresponding assets."

Data governance requires a larger engagement of the organization, always based on three main dimensions (people, technology, and processes) to put any aspect of data governance (knowledge areas) in place.

- **Roles and responsibilities (people)**: It's related to responsibilities regarding human aspects, such as legal definitions, ethical trade-offs, and social and economic considerations. It's also related to the roles of professionals focused on data management, which consists of the following.

- **Data owner** (also known as business owner): Accountable for the data of a specific process/domain. This role is responsible for data definitions, data quality, classifications, usability purposes, and so on.

  - **Data steward**: This professional ensures that the data policies and solutions comply with the definitions of data standards.

---

[8] Piethein Strengholt (2020), *Data Management at Scale.* O'Reilly Media, p. 185.

- **Data creator**: Who creates the data in compliance with the data owner's definitions.

- **Data user**: Who uses the data for its own needs and is accountable for setting requirements. Unlike the data consumer role, this role also collaborates on the definitions of requirements created by the data owner.

- **Data consumer**: Who uses data as intended by the data owner or user.

- **Application owner** (also known as data custodian): Maintains the core of the application (business delivery, services, application information, and access control) and its interfaces.

---

**Note**    It's important to mention that for some organizations, the data user and data consumer roles may converge into one. Similarly, data stewards and data creators roles may be also converged.

---

- **Organization and culture (people)**: It's related to the culture for the data management responsibilities and the behavior of the main business focus company's roles and responsibilities with the ones focused on data management.

- **Techniques and activities (process)**: It's related to how the data must be controlled, audited, and monitored. It includes data quality, data security (the two most critical), master data management, data life cycle management, business intelligence and analytics, data models, and data definition maintenance.

- **Tools (technology)**: It's related to the standardizations of tools and frameworks that allow data governance to be conformed to.

- **Data deliverables (technology)**: It's related to the data itself, such as its metadata, definitions, lineage, and every characteristic.

To manage each aspect of the data, the DAMA-DMBoK has defined 11 knowledge areas.

- **Data governance** provides guidelines and supervision for data management by setting up a system of decision rights over data per the enterprise's needs.

- **Data architecture** defines the model for managing data assets and establishing strategic data requirements and designs to meet them (e.g., strategic preparation of products, services, and data evolution to take advantage of business opportunities in emerging technologies.

- **Data modeling and design** is the process of discovering, analyzing, representing, and communicating data requirements in a precise form called the data model. Each model contains groups of components such as entities/facts/dimensions, relationships, keys, and attributes. Once built and approved, it needs to be maintained.

- **Data storage and operations** are the design, implementation, and support of data storage to maximize its value while counting on the operations to have support all over the data life cycle from planning to disposal of data.

- **Data security** ensures that confidentiality and data privacy are applied under regulations, contractual agreements, and business requirements. It involves the planning and executing procedures and security policies to provide proper authentication, authorization, access, and auditing of data and information assets.

- **Data integration and interoperability** include processes related to the movement and consolidation of data within and between data stores, applications, and organizations with interoperability, which means that communication happens among multiple systems (e.g., data sharing between applications and across organizations or ingesting external data.

- **Document and content management** is related to the management (planning, execution, and controls to capture, store, access, and use data) of the data life cycle documentation, usually needed to support legal and regulatory compliance requirements.

- **Reference and master data** (a.k.a. "data source of truth") is the recurrent reconciliation and maintenance of core and critical data for the business to ensure accurate, timely, and relevant data for consistent use across business areas, processes, and systems (e.g., customer lists, geographic location codes, and other data used to run the business.

- **Data warehousing and business intelligence** are related to the management (planning, design, and implementation) of a consistent analytical repository (data warehouse) that enables the enterprise to use

11

its data in a planned and monitored manner to make better decisions via analysis and reporting and create organizational value upon it (business intelligence).

- **Metadata** (a.k.a. "data about data") is related to the management process (planning, implementation, and control activities) to enable access to high-quality, integrated data regarding technical and business processes (e.g., data rules, data models, data flows, logical and physical data structures, business process representations, technology infrastructure, etc.)

- **Data quality** is the management (planning, implementation, and monitoring) of techniques to measure, assess, and improve the quality and fitness of data for use within an organization. In short, ensure the data is reliable and trustworthy: high-quality.

## Data Issues

*Because no organization has perfect business processes, perfect technical processes, or perfect data management practices, all organizations experience problems related to the quality of their data. Organizations that formally manage the quality of data have fewer problems than those that leave data quality to chance.*

—DAMA International

Data has a wide range of areas that must be considered to manage data efficiently. All of them will result in one important aspect: data quality.

You've seen that this data governance pillar is one of the most important aspects of the business because it is directly related to the data solutions created to address business needs and problems. In our current era, even more so, as it is fundamental to directly impact the business results generated through a technological revolution desired by any company that wants to keep existing: AI.

This chapter explains data products and data as a product and their purposes.

For now, let's keep in mind the importance of this key aspect of data: quality!

Conceptually and based on the numbers of surveys conducted by consolidated entities, data quality is pretty important to businesses, and the real impacts they can generate on the business are for good or bad. But what are the characteristics of poor data?

The poor data has at least one of these characteristics.

- **Non-compliant data definition (business + format)** is related to data that doesn't have a common conceptual understanding definition, or it's not aligned with the actual processes and expectations of the business definition along the company (e.g., define concepts and formulas calculations of revenues, losses, costs or efficiency for business units, currency representations for multinational companies, regulatory databases used for reports not respecting mandatory format fields).

- **Inaccurate data** is related to data that is not trustable due to not being reconciled with its domain's definitions in any aspect, such as its operational data sources, business/domain's rules definitions, regulations, and the reality of the business.

- **Duplicate data** is related to broken data regarding uniqueness rules. They generate wrong lower or greater results for measures of the business, which are used to compose important key performance indicators (KPIs) to monitor or to run operations for the business (e.g., duplicated rows for transaction events that wrongly increase measures results that calculate revenues, losses, costs, efficiency or risks.)

- **Incomplete data** is related to incomplete data due to missings, unavailability, or inaccessibility Should the missed data be treated with some statistical technique (such as median), or should it be kept for analytics and advanced analytics purposes? Problematic missed snapshot data is used to monitor the operation or to generate scores throughout ML models with up-to-date data for every hour, date, week, or month. From a managerial view, define how to account for the results of the business unit (revenues, losses, and/or costs) of closing months by determining whether it should append late events (such as sales) in the past months and modifying the historical.

All these characteristics apply to data used for direct core business solutions and internal solutions.

In terms of business impacts, data with low quality generates at least four types of negative impacts.

- **Regulatory violations and fines** are the result of reporting wrong data to regulatory or financial agencies, authorities, or customers. A business can lose benefits, customer confidence, or have their license to operate revoked.

- **Higher operational costs** happen due to unfit processes (directly or indirectly to the core business) using inefficient and non-standardized data architectures (e.g., manual processes using manual/ad hoc databases).

- **Loss of (opportunities to generate) revenue** from not understanding the market and internal opportunities to generate revenues, reduce costs and losses, improve the efficiency and quality of solutions, reduce risks for the business, and especially efforts at competitive differentiation in business strategy.

- **Negative reputational impact** from inaccurate official reports done or processes and decisions made based on poor data (e.g., increased customer service calls and decreased ability to resolve them, such as the Uber case in 2017).[9]

Organizations don't want to have any of these issues.

Therefore, we need solutions to mitigate the existing issues or their risks. In other words, we need data products to solve problems!

# Data Product Management, Data Management, and Data Governance

How are data management and data governance related to data products?

The organizations need and want data to perform regardless of the business sector.

- Business strategy, especially efforts at competitive differentiation

- Ongoing operations

- Business policies

- Regulatory and financial reportings

---

[9] Biz Carson (2017), "Uber is paying its NYC drivers 'tens of millions' because of an accounting error that underpaid them for years." *Business Insider.* Available at https://www.businessinsider.com/uber-paying-nyc-drivers-tens-of-millions-accounting-error-2017-5

The data must be managed by ensuring standards are met to enable the company to have it. In other words, companies must manage data through data governance definitions, which include the following:

- Understanding, supporting, and evolving the information needs of the enterprise and its stakeholders, including customers, employees, and business partners

- Capturing, storing, processing, protecting, and ensuring the integrity of data

- Ensuring the quality of data

- Ensuring the privacy and confidentiality of data in compliance with data security and law definitions

- Preventing unauthorized or inappropriate access, manipulation, or use of data and information in accordance with laws of data

- Ensuring data can be effectively used to enable additional value to the enterprise

Besides the solutions directly developed for the core business, such as ML models to score customers' risk or suggest products for customers, each data management responsibility has great opportunities in terms of solutions' implementations to make the process fast-paced and smooth.

Furthermore, data quality is one of the most fundamental pillars of AI and analytics solutions.

Hence, data products, data management, and data governance are completely related and must be considered constantly.

Please remember that, especially if you are a leader, No analytical solution, be it AI or the traditional BI (business intelligence), fully results in the best impact potential without quality and governance. Actually, it can result in the opposite. When talking about data strategy, the three pillars must be considered together!

# Examples of Opportunities Among the Three Pillars

Examples are always a great way to understand concepts and their importance!

Therefore, let's talk about some examples of potential opportunities to create products to address data issues or address issues with data.

## Core Business Needs

- Automate score calculation of the potential risk of a customer to offer products or not

- Automated suggestion of products based on customer's interests and experiences

- Automated prediction of the volume of sales, potential revenue, costs, losses, and profits

- And many other scenarios

## Data Management Needs

- **Automated data wrangling:**[10] Identifying and fixing errors, filling missing values, and standardizing data formats

- **Data discovery**: Enabling automated relationships information and other recurrent necessary evaluations to reduce inefficiency due to manual data analysis

- **Automated anomaly detection**: Identifying issues early on and preventing them from becoming bigger problems

---

[10] Data wrangling is the process of transforming raw data into usable data, by discovering, structuring, cleansing, enriching, validating and publishing.

# More Color: Data Product Management

In 2017, I got my first experience with data product management. I assumed the data product ownership of a fintech company that was growing exponentially.

I need to say the company was pretty innovative because it was a pretty new role that didn't exist in marketing yet. At least we couldn't find any information on the market at that time. Thus, we created this role, called that time as data product owner (based on scrum roles), to address the necessity of treating data as a product.

It was the first time a team approached the technical solutions as products. After some time, other teams decided to leverage the rationale, and infrastructure products emerged within some teams.

The context my team and I were facing was that we were a central data team supporting the whole company, and we could not escalate the solution's implementation by on-demand requests anymore as soon as the company grew. Furthermore, we had many other issues related to the fact we were not treating it as a product.

- **Business context and prioritization for business impact**: The interaction with stakeholders was limited, impacting the business knowledge and context to prioritize initiatives bringing clear value and results.

- **Communication and data strategy**: Many stakeholders, purposes, and projects with a decentralized strategy generate many mismatches in communication.

- **Agility and interoperability:**[11] As the backlog was long and without a strategy, we lost the agility to interact with the stakeholders about the solutions on time and ensure the data has the interoperability necessary for the business with updated and transparent integrations as soon as new business rules or services emerge.

- **Data architecture**: We had a legacy platform, which we were innovating by migrating to a new one. However, both should be maintained. And the results of the new stack and its impacts were constantly requested.

- **Project management**: Many projects were necessary to address all the existing and emerging needs for the platform and the analytics solutions. One centralized view of data strategy becomes necessary.

Data governance started to pop up from paper to request actions due to data privacy laws, such as the LGPD (Brazilian data privacy law). It was time to work strategically by considering our goals for the platform and the business core goals.

At that time, no common definitions of technical (and internal) product management existed. By the way, I was the first product owner (now usually called a product manager) working with platform/internal products.

I created my own strategy to manage data products using agile frameworks for product and project management. I used my knowledge and experience in data analytics to design the solutions with the teams. I aligned internally to have two separate products: the platform and data for analytics purposes. Maybe I should have written a book that time!

---

[11] Interoperability it the ability of systems, applications, products or any other entity to exchange and use information (connect and communicate) with other entities in a coordinated way, by avoiding special effort from the end user.

Two years later, data mesh, which is discussed in this chapter, started to become popular. This new concept was interested in creating a path to convey similar needs and responsibilities.

# Data Mesh

> *Data mesh is a decentralized sociotechnical approach to share, access, and manage analytical data in complex and large-scale environments—within or across organizations.*
>
> *Data mesh is a new approach in sourcing, managing, and accessing data for analytical use cases at scale.*
>
> —Zhamak Dehghani, *Data Mesh*, 2022[12]

The data mesh concept was globally recognized in 2019 when Zhamak Dehghani wrote a blog article[13] on this innovative vision to manage data by scaling the data-driven approach.

Due to strong interest from the data community, Zhamak wrote a book in 2022 that explains the data mesh approach.

A data mesh is based on four pillars.

# Domain-oriented Ownership

Domain-oriented ownership is the concept of decentralization of data management, as well as the business representation of it within the domain.

---

[12] Zhamak Dehghani (2022), *Data Mesh: Delivering Data-Driven Value at Scale*, O'Reilly Media.

[13] Zhamak Dehghani (2019), "How to Move Beyond a Monolithic Data Lake to a Distributed Data Mesh," MartinFowler.com. Available at https://martinfowler.com/articles/data-monolith-to-mesh.html.

In other words, this pillar expects the following.

- **Representation of the business**: The detailed domain represents the core business context and goals.

- **Decentralized ownership**: The business domain area is the group responsible for the data implementation and ongoing[14] data management instead of having a central data team for the company. It's expected to ensure the accuracy and timeliness of the data.

What are the benefits for the business?

- **Faster data analytics adaptations necessary to the business**: As soon as new products or features emerge on that domain (generating new services and data operational sources) or are deprecated (adjustments or deprecation of data sources), the domain's team can implement the necessary analytical adjustments and improvements. It'll support the business with timely data for operational monitoring and decisions through machine learning engineering (MLE) with up-to-date business rules.

- **Trustful data source for analytics purposes**: Risk removed or at least very well mitigated of mismatched data for the domain among business areas with different concepts due to the ownership of the business domain defined to the appropriate responsible.

- **Operational efficiency for implementation and ongoing**: The setup of ownership of the data to the business domain org reduces the dependency of teams

---

[14] Ongoing is related to continuous work, which englobes the maintenance and evolution of data analytics, software, or applications.

by having fewer teams involved, fewer data pipeline efforts and dependencies, and more efficient data for analytics and advanced analytics.

# Data as a Product

Data as a product is a paradigm definition where the data is a product with autonomy and its own life cycle.

It is especially necessary in the data mesh approach due to the definition of domain-oriented data.

Data as a product has eight characteristics in the data mesh architectural framework.

- Discoverable

- Addressable

- Understandable

- Trustworthy and truthful

- Natively accessible

- Interoperable and composable

- Valuable on its own

- Secure

These definitions are also considered in this book, which is explained in more detail in the next section, together with other definitions.

What are the benefits for the business?

- **Culture of data as a product**: Change the dynamic of work of the professionals with the data, by having customers consume it, instead of implementing and multiplying silos.

- **Usability of data with low friction**: Define business context by domain and data evolution and maintenance, autonomously managing dependencies, and data sharing with explicit contracts for usability.

- **Cost efficiency for implementation, ongoing, and innovation**: Implement business solutions by using high-quality data.

## Self-Serve Data Platform

The self-serve data platform is related to empowering domain teams by developing features and services that address data integration across cross teams by managing the whole life cycle of data products, from building to maintenance.

It's focused on reducing the friction of data usability by having solutions for each data profile: manager and user. Such as data management and security: data sharing from the end-to-end journey, ensuring reliability, and experience of discoverability and usability of the data products.

What are the benefits for the business?

- **Automation for security and compliance**: Services and solutions to ensure standards for all data products

- **Less complex data management effort**: Due to automation and complexities, abstractions to manage the full life cycle of data products

- **Reduction of cost**: Fewer professionals performing complex work with required specialized knowledge

## Federated Computational Governance

The goal for federated computational governance is to ensure a federated (a.k.a. enterprise-level) accountability and decision-making structure, balancing autonomy and agility while ensuring integration along the company. The standards are defined by a group of expert representatives of domain, platform, legal, security, and so on.

What are the benefits for the business?

- **Federated definitions** due to enterprise standards for the company, leveraging the different perspectives for data needs

- **Integration of solutions** due to federated definitions ensures consistency and integrations among different areas with common solutions/patterns

- **Faster resolutions** due to the automation to apply integration among the areas

How do the pillars interact?

- **"Domain-oriented ownership" to "data as a product"**: This relationship prevents data from being siloed, as each domain owner is responsible for building the source of truth data and sharing it across the company with the data users.

- **"Domain-oriented ownership" to "self-serve data platform"**: This relationship empowers domain teams to implement the platform solution necessary for their business operation.

- **"Data as a product" to "self-serve data platform"**: This relationship utilizes the data as product ownership (creation, maintenance, and ongoing) of new and leftover old systems by enabling domain teams to use and configure data platforms as necessary for the business unit contexts.

- **"Domain-oriented ownership" to "federated computational governance"**: This relationship makes it easier for domain teams to govern the data according to standards resulting from federated computational governance.

- **"Data as a product" to "federated computational governance"**: This relationship enables the development of new data as product solutions and the creation of great value for the business by leveraging the federated computation governance results, such as the possibility to integrate with other data as a product through standardized definitions.

- **"Federated computational governance" to "self-serve data platform"**: This relationship helps ensure the policy enforcement of federated governance definitions through the data platform design created for the org by domain teams.

# Data Products vs. Data as a Product

Although the majority of the market defines Data Products and Data as a Product with converged definitions, we'll perform a different approach in this book. We'll talk about them distinctly, clearly bordering them, as I see it makes more sense. Data products and data as a product have different characteristics and goals. Before, ignoring a little bit the distinctions, it's important to have clear the similarities. Both fit to the product's definitions.

- They solve concrete issues and are based on that.

- They have a clear and specific audience, which includes customers and stakeholders.

- They demand to be supported via maintenance and ongoing, as business processes are related to them.

- They demand clear ownership and sponsorship.

Regardless of the type of product, as mentioned in a 2020 *Harvard Business Review* article,[15] it's fundamental to apply key product development principles. It means identifying the main problems and, based on that, addressing them with agility, iterability, and reusability. That being said, let's discuss about the differences between them.

Table 1-1 describes the differences between a data product and data as a product.

***Table 1-1.*** *Definitions and Differences Between a Data Product and Data as a Product*

|  | DATA PRODUCT | DATA AS A PRODUCT |
| --- | --- | --- |
| **PURPOSE** | Boost data management by implementing solutions and developing data as a product or fundamentally using data as part of a solution with a specific analytical or business goal. | Use the data itself for analytical, business, or decision-making purposes. |
| **CUSTOMERS** | • Data producers<br>• Data consumers | • Data consumers |
| **IS IT A PLATFORM?** | Generally, yes, but not all products are platforms by itself. However, they should be an important part of it. | Not by itself, but it's an important part of it. |

*(continued)*

---

[15] Jedd David, Dave Nussbaum, and Kevin Troyanos (2020), "Approach Your Data with a Product Mindset," *Harvard Business Review*. Available at https://hbr.org/2020/05/approach-your-data-with-a-product-mindset

***Table 1-1.*** (*continued*)

| | DATA PRODUCT | DATA AS A PRODUCT |
|---|---|---|
| **OWNERSHIP ROLE** | • Platform teams: domain ownership teams or federated computational teams (potential centralized teams defined by the company to attend to company-level tools needs)<br><br>• Domain Owner teams (a.k.a. business owner teams): to define solutions for their specific business needs. | Domain owner teams (a.k.a. business owner teams) |
| **EXAMPLE OF PRODUCT** | Data management<br>• Ingestion solution<br>• Data security solution<br>Implement data as a product<br>• Data architecture frameworks<br>Data is the foundation for a business goal<br>• Implement frameworks (a.k.a policies) for automated business decisions<br>• Engines to search and rank data | Domain and subdomain data<br>Actual, predicted, or classified data for<br>• People and hiring (inside the HR business subdomain)<br>• Credit underwriting (inside the Credit business subdomain)<br>• Suspects and fraudsters (inside the Fraud subdomain)<br>• Victims of fraud (inside the Fraud subdomain) |

(*continued*)

***Table 1-1.*** *(continued)*

| | DATA PRODUCT | DATA AS A PRODUCT |
|---|---|---|
| **MAIN BENEFITS** | • Operational efficiency due to automation<br>• Cost efficiency due to better cost management and operational efficiency | • Data quality and trustworthiness for analytical needs<br>• Operational efficiency due to a clear source of truth<br>• Cost efficiency due to redundant work removed<br>• Clear ownership for maintenance of data |
| **MAIN CHALLENGES** | • Prone to depriorization versus direct business initiatives: except the products that are foundation for a business goal, the data products are usually less valued by senior leadership due to indirect results delivered for the business core operation.<br>• When escalated can easily become a centralized and bureaucratic technical product | • Definition of ownership and business rules for subdomains with full or partial cross characteristics.<br>• Difficult to measure and prove the result by itself for the business during the development. It usually requests a bad scenario to prove the post result or a business / analytical goal to relate the results of it |

# What Is a Data Product?

A data product is a solution (software or framework) that boosts data management or develops data as a product.

Regardless of the goal, it usually has two types of customers.

- **Data producers** own data and ideally produce it as a product by sharing it with customers.

- **Data consumers** consume the data produced for specific purposes.

To get more color about the concepts, let's evaluate some examples! Table 1-2 provides examples of data products that boost data management.

***Table 1-2.*** *Data Products That Boost Data Management*

**TYPE OF PRODUCT: TO BOOST DATA MANAGEMENT**

| | | |
|---|---|---|
| **EXAMPLE CASE** | Data Ingestion Solution | Data Access Solution |
| **PURPOSE** | Create solutions to automate the ingestion from common sources (e.g., Stitch, APIs, or in-house solutions that apply contract best practices and enterprise federated definitions to enable standard data ingestion. | Automated solutions to apply federated security access to data for users of them, such as access to schemas, tables, attributes, or by population of data (as not PII data, or for geolocation, or a specific product, etc.). |
| **CUSTOMER** | Data producers<br><br>• Analytics engineers treat the data to create data as a product<br><br>• SW engineers that consume that to integrate with other operational services | Data producers: The data owners and stewards of data who define who can access which data and for which purposes<br><br>Data consumers: Those who need to consume the data for specific purposes; for example, risk analysts for regulatory purposes or finance analysts for accounting management |

Table 1-3 provides examples of data products that implement data as a product.

***Table 1-3.*** *Data Products That Implement Data as a Product*

| TYPE OF PRODUCT: IMPLEMENT DATA AS A PRODUCT |
| --- |

| | |
| --- | --- |
| **EXAMPLE CASE** | Framework engines that treat data to create data features |
| **PURPOSE** | Have a standard framework that enables the organization to create features for their own needs or eventually share and consume from others. The solution is responsible for the process of generating data as a product. |
| **CUSTOMER** | Producers: Data owners and stewards who create products for their own domain<br>Data consumers: Those who create data as a product for their domains based on their own data or from other domains |

# Data as a Product

Although I respect Zhamak's vision and definition that not every data is a product, I see that every data is a product by itself, regardless of the medal of the data, as soon as:

- They solve concrete issues and are based on that. At least, they should; otherwise, the investment in its creation will just generate costs, without reasons for it. Actually, it wouldn't be an investment, but waste of money.

- They have a clear and specific audience, such as customers and stakeholders, even if used for an ad hoc scenario or to create part of a data pipeline.

- They demand to be supported via maintenance and
  ongoing maintenance. Even if the business processes
  are related to them, they generate costs and consume
  storage and computing capacity. Thus, it must be
  maintained and deeply managed, even if it means to
  conclude its life cycle.

- They demand clear ownership and sponsorship.

See the "Medallion Data Architecture" section in Chapter 3 to learn
more about medalized data.

Data is a product by itself due to its core capacity to solve issues
through its fundamental usability for analytical, business, or decision-
making purposes.

It has customers (usually, but not limited to, business analysts, data
scientists, data engineers, and others).

Its main goal is to make the data usability simple. As Zhamak[16]
suggested (2022), the data mesh concept defines eight characteristics of
data as a product.

- **Valuable**: I'd say that value is the main characteristic of
  data as a product. It has and delivers value on its own.
  I agree with Zhamak's description of this characteristic
  that if there's no value, it shouldn't exist. I'd rephrase it:
  if it's not solving an issue, it shouldn't exist.

- **Trustful**: Truthfulness is one of the most important
  characteristics of data as a product. As the domain
  source of truth the whole company uses for different
  business needs, it's fundamental to trust the data. Data
  as a product must comply with the domain's business
  context and needs and adhere to actual and predicted

---

[16] Zhamak Dehghani (2022), *Data Mesh: Delivering Data-Driven Value at Scale,*
O'Reilly Media.

data by ensuring automated integrity checks, data
lineage, and reliability. After all, reliability is related
to trust, and the data must be available. Everything is
not necessary only due to the creation and early usage
stages but also an equally important post-adoption
stage—for instance, postmortem and crash evaluations.

- **Understandable**: The data as a product must
  be understandable for any user. The users must
  understand the domain's context through the data
  structure and population. Also, they should be able
  to understand the operation of the data through its
  semantic data. As stated by Zhamak, a self-serve
  method of understanding is a baseline usability
  characteristic.

- **Discoverable**: Data as a product has and automatically
  provides full information about itself during its journey.
  It's searchable by its customers and curated by the
  governance team. It contains datasets' information,
  ownership, top use cases, and applications enabled.

- **Addressable**: Data as a product must offer customers
  a permanent and unique address. It considers the
  data domain's characteristics as very important to
  its customers' needs of usability: ad hoc and manual
  access or programmatically. Moreover, the availability
  of the data is critical, requesting the data owner to keep
  the data available and reliable, even considering the
  expected ongoing growth of the business domain. It
  also has to have documentation with all information
  related to these topics.

- **Secure**: Data as a product has strong security controls, where data is shared adequately with data users and controlled by data owners. The security aspects entail versioning and automated testing of policies, encryption definitions, regulations and laws rules, and so on.

- **Interoperable**: Data as a product is available and integrates with multiple domains. It has a global definition for its own domain but harmonizes with all definitions that would be different in other moments by other domains.

- **Natively accessible**: A data as a product is accessible and readable by all data users as its personas native modes requests (e.g., data analysts who are comfortable with exploring data in spreadsheets, others with query languages, MLEs who consume data frames to train their ML models, etc.)

In 2018, Simon O'Regan published an article[17] defining the five types of data products: raw data, derived data, algorithms, decision support, and automated decision-making.

I'd adjust this definition for the five possible classifications of data as a product. It is mainly because data as a product can solve issues, has customers, and requests maintenance. In short, it's a product by itself.

---

[17] Simon O'Regan (2018), "Designing Data Products," towardsdatascience.com. Available at https://towardsdatascience.com/how-to-quickly-anonymize-personal-names-in-python-6e78115a125b/

- **Raw data:** The goal is to have raw data for analytical or business purposes. The data is produced in the operational environment and is shared with consumers in its original format (without treatment) or most original format (with a few pre-processing/treatments, such as putting it in a tabular view).

- **Derived data:** The goal is to have canonical data for analytical and business purposes. The data is produced based on the raw data. It is wrangled in the analytical environment by applying cleansing, treatments, and business rules in adherence to its domain's definitions. Then, the final data produced is shared for consumption.

- **Algorithms:** The goal is to predict or classify the data for a specific business goal. The data is produced upon derived and raw data, with an algorithm receiving some data as parameters and outputting (serving the data for consuming) information or insight. They can be algorithms-as-a-service, usually available through API, or used ad hoc.

- **Decision support:** The goal is to use data for decision-making. The data is produced and shared with the information provided for user decision-making. Another adjustment that I'd apply from Simon's definition is that moreover sharing the data for analytics and decisions, this type of data as a product may also apply some intelligence to suggest solutions, performed by algorithms. However, the product is still not responsible to take it by itself. It's the responsibility of the customer.

- **Automated decision-making**: It's another adjustment I'd make from the definition suggested in 2018. The data produced is a decision already made instead of a suggestion. Algorithms perform the intelligence, and the data as a product is shared for consumption with the final decision made. The decision-making can change and adapt itself based on additional data consumed by the algorithm through the customer's journey usage of the product.

The delivery interfaces can be APIs, dashboards, and web elements (e.g., scroll of products, audio, videos, etc.).

Table 1-4 provides data as a product definitions and examples.

*Table 1-4.  Data as a Product Definitions*

| DATA AS A PRODUCT TYPES | | | | | |
|---|---|---|---|---|---|
| | **RAW DATA** | **DERIVED DATA** | **ALGORITHM** | **DECISION SUPPORT** | **AUTOMATED DECISION-MAKING** |
| **GOAL** | The goal is to have raw data for analytical or business purposes. | The goal is to have canonical data for analytical and business purposes. | The goal is to predict or classify the data for a specific business goal. | The goal is to use data for decision-making. | The goal is to make decisions without human interaction for business purposes. |
| **DEFINITION** | Data produced in the operational environment and shared for consumers in its original format | Data produced upon the raw data wrangled in the analytical environment, and final data produced shared for consumption | Data outputting information or insight via an algorithm | Data produced and shared with information provided for decision-making by the user | Decisions adapted by the algorithm, based on additional data consumed through the customer's journey usage of the enterprise product |
| **EXAMPLE CASE** | Server logs data | Fraudster and suspect data | Algorithm-as-a-service for risk assessment | Customized and automated suggestions for products to buy or use | Customized and automated decision-making for content presentation |

*(continued)*

*Table 1-4.* (*continued*)

**DATA AS A PRODUCT TYPES**

| | RAW DATA | DERIVED DATA | ALGORITHM | DECISION SUPPORT | AUTOMATED DECISION-MAKING |
|---|---|---|---|---|---|
| **PURPOSE** | Based on raw data, able to evaluate the accesses to a web server to evaluate the requester IP and status HTTP | Given a list of fraudsters and suspects, define policies to block offers or usability of products | Given a prospect, define the level of risk to offer credit to the person | Present a list of potential products to be bought or used by the customer based on their historical preferences | Present the content for consuming through written (e.g., Medium), image (e.g., Instagram), video (e.g., TikTok), or other formats based on decisions made by algorithms, which use the historical preferences of the user for it |
| **CUSTOMER** | Consumers of the raw data, which in this example is the tech team responsible for the availability and specific service | Consumers of the canonical data created, which in this example are the business teams responsible for defining the policies for their products | Consumer of the assessment realized by the algorithm, which in this case is the credit underwriting org | Consumers of the suggestions of decision-making (e.g., Amazon customers, Netflix customers, and so on) | Consumers of the final decision made (e.g., consumers of the content's platform |

37

# Are Data Products and Data as Product Platforms?

A platform is a product that serves or enables other products or services, which usually serve consumers through APIs but may offer other mechanisms for consumption (Gartner).[18]

Although not completely new, as MIT published an article about platforms in 1998, recently, it has become a trend to modernize enterprise software delivery, particularly for digital transformation.

The mission of platforms is to accelerate the delivery of applications and the pace at which they produce business value. The main reason is the improvement of productivity through self-service capabilities with automated infrastructure.

Following these concept definitions, you can say both are related to platforms but in different depths.

- A **data product** can be a platform by itself, and if it's not, it is an important part of a platform as it is the means for the final resolution of a given problem. A clear example is the framework case, which addresses goals directly related to the main enterprise's business strategy.

- **Data as a product** is part of a platform but not by itself. The reason for this is that without data, nothing can be solved, period. Its definition states that the data is necessary for analytical, business, or decision-making solutions and addresses the needs by themselves. However, it requires detailed effort to implement them individually, and it is built and performed upon platforms as a requirement. An example of this is

---

[18] Gartner.com, Glossary: (Platform Digital Business). Available at https://www.gartner.com/en/information-technology/glossary/platform-digital-business

product recommendations on Amazon's site. A site is necessary in the background to enable the product's usability: recommendation of products.

# 1.3.  Skills and Techniques to Manage a Data Product

Gartner (2023)[19] says that by 2026, more than 80% of enterprises have used generative AI APIs and models and/or deployed GenAI-enabled applications in production environments, up from less than 5% in 2023.

We're in the early stages of the AI era, which has been evolving exponentially. If companies want to remain active, it's fundamental to start this practice.

However, every time trends emerge, the same behavior and affirmations occur from professionals along the market: "Let's forget everything lived so far. This new technology is the solution to everything. The rest is obsolete!"

This common affirmation from professionals from the market, including many senior ones, does not make any sense! AI is bringing changes in the market, as technology evolution has been changing it and the way we live with them, since always. However, to move forward with this important and revolutionary technology's range of tools available in AI, it's necessary to ground its base: data. Even if the plan is to evolve in a way that we'll use AI for that, which yes, we'll do in mid-long term. But ignoring the past and evolution's history is a huge mistake, especially talking about its reasons and challenges. The reason is that the same that occurs for human civilization history, occurs in technology: the challenges

---

[19] Gartner.com (2023), "Gartner Says More Than 80% of Enterprises Will Have Used Generative AI APIs or Deployed Generative AI-Enabled Applications by 2026." Available at https://www.gartner.com/en/newsroom/press-releases/2023-10-11-gartner-says-more-than-80-percent-of-enterprises-will-have-used-generative-ai-apis-or-deployed-generative-ai-enabled-applications-by-2026

tend to be similar and usually cyclical, but in different depths during its journey. That's why it is so important to understand the path journey performed of everything, to get more clarity about the possible scenarios to be faced, and knowledge learned that might be helpful for current or future similar contexts and scenarios.

This evolution process is discussed in Chapter 2.

Before that, let's foment important fundamental knowledge to manage data products and data as a product!

They are based on three main topics: data and analytics, product management, and project management.

# Data, Analytics and AI Definitions

## Operational and Analytical Data Architecture

Chapter 2 goes deeper into analytical architecture. For now, the most important thing is to have clarity about the operational and analytical architectures, why they exist, and what they solve.

Operational architecture, or online transaction processing (OLTP) systems, is the platform where the company's events occur in real time in a transactional environment. The repositories are used by services and microservices and are responsible for performing CRUD (create, update, delete) operations on them, ensuring efficiency in writing and reading by rows. They must usually be reliable and available.

One example of an operational process is a purchase made for an item in a store. For it to happen, the operational architecture must be able to, in seconds, integrate among the bank used by the customer's credit card, the merchant's acquirer, where the customer is making a purchase, and the credit card brand, the credit card regulator, partner of the customer's CC bank to have a transaction approved.

As you can imagine, it is a long process with much interoperability, and it must happen quickly, with availability.

For this reason, it requests a focused and efficient architecture for its mission.

Once operational data is available, it's necessary to get the data originated in the operational environment to evaluate results. But how should we do it if analytics requests cannot impact the operational architecture on the datasets designed for operational needs? An apart environment emerged to address it: the analytical environment, also known as online analytical processing (OLAP).

Its main goal is to enable the creation and storage of analytical information that supports business exploration, discoveries, reports, and the creation of algorithms for business purposes.

In short, the analytics and advanced analytics live in analytical architecture.

As you'll see during our journey in this book, due to the evolution of technology, there is now more integration between them and in a cycling manner. However, they still have distinct missions.

## Analytics and Advanced Analytics

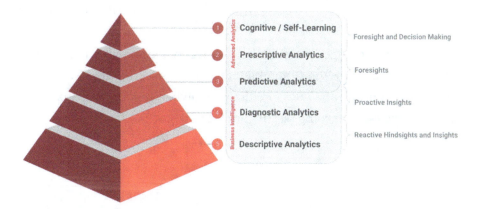

*Figure 1-1.* *Analytics and advanced analytics*

# Business Intelligence

Ralph Kimball (2002),[20] a significant contributor to the design and evolution of analytical architecture, depicts business intelligence as a generic term for leveraging the organization's internal and external information assets to make better business decisions.

Business intelligence, also called analytics, is related to the information generated by looking at and analyzing data. It is a term that englobes infrastructure, applications, and best practices to analyze information for businesses to make optimized decisions.

In terms of applications, it provides features for OLAP during the whole data journey, focused on analysis purposes, delivery of information through reports and dashboards, and platform integration with internal and external solutions for different objectives.

*Analytics brings information and information capabilities into the world of business, where such information has never before been possible. And there are as many forms of analytics as there are stars in the sky.*

—Bill Inmon (father of the data warehouse)

There are five stages for analytics, defined by the analytics maturity matrix (used to assess the companies' analytics stage): descriptive analytics, diagnostic analytics, predictive analytics, prescriptive analytics, and cognitive/self-learning (see Figure 1-1).

---

[20] Ralph Kimball and Margy Ross (2002), *The Data Warehouse Toolkit: The complete Guide to Dimensional Modeling*, John Wiley & Sons, Inc.

- Descriptive and diagnostic analytics are related to the traditional analytics/ business intelligence process.

- Predictive analytics is an intersection between the traditional analytics/ business intelligence process and the advanced analytics process.

- Prescriptive analytics and cognitive/self-learning stages are related to advanced analytics processes.

## Advanced Analytics

Advanced analytics is related to sophisticated techniques that go further than traditional business intelligence for business needs and purposes. It is the practice of applying AI techniques.

## Stages of Analytics

- **Descriptive analytics**: This stage is used for reactive hindsight and insights by answering questions like What happened? or What is happening? These analytics are usually performed during data explorations for different purposes, such as evaluation of the specific customers' behaviors regarding specific products, process results evaluation, benchmark of products or processes, and so on. The analyses are commonly performed through traditional BI tools, such as DataViz, ad hoc analytics tools, and the famous and eternal Microsoft Excel with its spreadsheets.

- **Diagnostic analytics**: This stage is used for reactive insights by answering why something happened and to start proactive insights by answering why something is currently happening. These analytics are usually performed by going further on the description analytics and producing insights that answer the past and present. The analyses are commonly performed through traditional BI tools, such as DataViz, ad hoc analytics tools, and Excel. Moreover, the analyses are done by using traditional BI techniques, such as drill-down in historical data (to identify and compare patterns), data discovery, correlations, and data mining, such as the application of clusterization models to identify patterns based on common characteristics and behaviors of the population evaluated.

- **Predictive analytics**: This stage is an intersection of traditional and advanced analytics. It is used to generate proactive insights, but mainly for foresight purposes, by predicting what is likely to happen. This analytics is performed by going further on the diagnostic analytics, and the analyses are commonly performed by applying advanced analytics for predictions for the future (e.g., regression analysis to predict specific scenarios, such as sales, churn, and turnover).

- **Prescriptive analytics**: this stage is the first full advanced analytics used for foresight. Once it is clear what happened and is happening, the reason for it, and the prediction of what will happen, the next step is to answer a business question: What should we do about it? It is a step further from prediction analytics. Good examples of this are algorithms of recommendations that make suggestions based on the preferences of the public target.

- **Cognitive/self-learning**: Cognitive analytics is the highest stage in advanced analytics. It is the evolution of prescriptive analytics, with a high level of automation and focus on decision-making, by emulating human thought processes. It is focused on answering questions related to insights into what's not yet known; for example, it can be used to evaluate new business opportunities by examining trends, behavior, and interests of existing customers to, upon that, guideline the business to tailor products and processes.

## Advanced Analytics

*One of the most promising forms of analytics is that of forward-looking analytics.*

—Bill Inmon

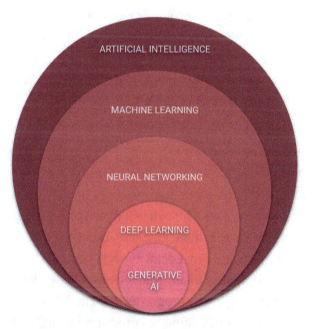

***Figure 1-2.*** *Narrow AI*

# Artificial Intelligence

Artificial intelligence (AI) is a multifaceted discipline within the larger field of computer science AI focuses on reproducing similar human competencies, such as creativity, learning from experiences, interpreting data, and making decisions equivalent to a human being in a machine.

The following describes the types of AI.

- **Narrow AI** (a.k.a. weak AI): It is the current stage of AI. The main goal is to be specialized with high proficiency in a specific area by operating on specific tasks and within limited constraints. Once well-trained, they perform even better than humans. However, it's incapable of generalizing its knowledge to execute activities outside of its specialty. Automating repetitive tasks such as image recognition or decision-making processes is very good. Other usage examples are personal assistants such as Apple's Siri or Amazon's Alexa and NLP (natural language processing) based solutions such as OpenAI GPT. Its performance depends on the data available for training.

- **Artificial general intelligence** (AGI/strong AI) is a human-like AI. We are in the very early stages of this era, still under ongoing research. It's the closest future expectation. The concept of artificial general intelligence is that it is capable of understanding, learning from minimal data, applying knowledge in multiple domains, in any intellectual task as a human, and improving itself autonomously. The "closest" examples to its definition are products created with generative AI, such as personal assistants and chat GPT, which create very structured answers to aleatory questions. Although even using great and innovative algorithms, they lack the key definition of AGI: genuinely understanding the reasons for its decision-making and capacity to do analyses and decision-making outside of its training area. Therefore, they are still narrow AI solutions. However, we're evolving to AGI, and the aim is to deeply understand and replicate the complexities of human intelligence.

- **Superintelligent AI** is a hypothetical AI with many speculations. Its concept is that it surpasses human intelligence. It would perform artificial general intelligence, including social skills like persuasion and negotiation, and incredible and unfeasible activities for humans, such as easily finding solutions for big humanity's challenges, such as unknown and complex diseases or the climate.

## Machine Learning

Machine learning is a branch of AI that allows machines to learn and improve from experiences through algorithm applications. It is focused on learning from data specifically and related techniques (see Figure 1-2). In other words, it operationalizes and "productizes" the data science models created.

Mariette Award and Rahul Khanna (2015) say that machine learning harmonizes the essential relationships between data and information by receiving a specific data input for analysis and outputting predicted future events or unknown scenarios to the computer.

For this, among other that are emerging, there are three main types of machine learning: supervised, unsupervised, and reinforced.

Let's take a look on the main types of machine learning.

- **Supervised**: This term comes from the process of teachers supervising and mentoring their students by providing detailed information and constant validations. Then, the students learn during a detailed process. This process is the reference for the AI approach: ML models that use and need data labeled to be used during the training stage, and once performed, the ML model can extrapolate the learning to new and unseen data. Table 1-5 highlights AI examples of supervised learning models.

***Table 1-5.*** *AI Examples of Supervised Learning Models*

| TYPE OF DATA | TYPE OF BUSINESS QUESTIONS | SOME EXAMPLES OF ALGORITHM | EXAMPLE OF USABILITY |
|---|---|---|---|
| **Regression** Numeric target to be predicted | How much will this customer use this product/service? Should we give credit for this prospect? How much will the infrastructure cost? | Linear regression Decision tree and random forest | Sales forecasting Risk assessment Pricing Budget forecasting |
| **Classification** Binary target to be predicted | Will this customer buy this service or customer after the marketing campaign? Which of the three products or services will this customer likely buy? Should the email be classified as spam? | Logistics regression K-nearest neighbors SVM (support vector machines) | Spam detection in emails Financial performance comparison |

- **Unsupervised learning**: This term comes from the process of teaching when the teachers don't detail evaluate the student, but they generate their own conclusions. The unsupervised learning algorithms learn from unlabeled data and identify patterns and relationships within the data. Table 1-6 highlights AI examples of unsupervised learning models.

***Table 1-6.*** *AI Examples of Unsupervised Learning Models*

| TYPE OF DATA | EXAMPLE OF BUSINESS NEEDS | TYPE OF ALGORITHMS | EXAMPLES OF USABILITY |
|---|---|---|---|
| **Clustering** Divide the database by similarities | What is the demographic analysis of our customers based on their financial behavior? Which customer's behaviors are common and uncommon? | K-means-clustering | Demographic analysis Customer segmentation Recommender systems Anomaly detection |
| **Association** Identify sequences | What is the audience's feeling about the brand on social networks? How can we check if the user is not a robot but a human being? How can we check out the new registers with an internal fraudsters control list? | Artificial neural networks | Sentiment analysis Face or image recognition |
| **Dimensionality reduction** | How can we reduce this big volume of features to a lower, but still efficient volume? | PCA (principal component analysis) | Features selection to design ML models |

- **Reinforcement learning**: The algorithms are based on rewards and penalties. Basically, the algorithm's performance improves by interacting and adjusting its actions based on the feedback received from the environment. Table 1-7 provides AI examples of reinforcement learning models.

***Table 1-7.*** *AI Examples of Reinforcement Learning Models*

| STUDY FIELD | EXAMPLE OF BUSINESS NEEDS | TYPE OF ALGORITHMS | EXAMPLES OF USABILITY |
|---|---|---|---|
| NLP - Natural Language Processing | Translate sentences from one language to another expected language, as necessary for the audience of it. Improve the productivity of the journalism team by saving their time to focus on complex themes and projects, instead of simple daily newspapers | Artificial neural networks Deep neural networking | Language translation Newspaper writing generation |
| Robotics | Industrial robots that manipulate objects using raw sensor inputs for logistics warehouses. Emulate an artificial player to play against other users (artificial or not). | Artificial neural networks Deep neural networking | A robot performing warehouse Simulate a game play |

## Artificial Neural Networks

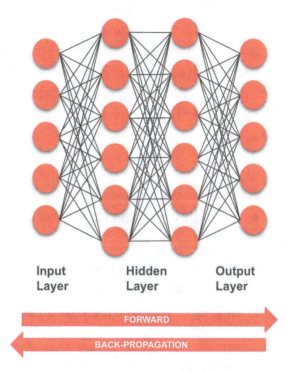

*Figure 1-3.*  *Artificial neural networks*

Artificial neural networks (ANNs) is a subset of machine learning, and its concept of functionality is inspired and mimics how the brain biologically works: neurons (nodes for ANNs) work together to identify and evaluate behaviors and scenarios for conclusions (see Figure 1-3).

51

ANNs are composed of the following.

- **Nodes** contain their own weight for the features[21] and a threshold result for the business requirement defined for the goal under evaluation. Whether the final output of the node is equal or greater to the threshold, the next layer's node is activated and receives the data from the previous one. Otherwise, the next layer's node doesn't receive any data.

- **Layers**

  - **Input layers** are responsible for receiving the external data and starting the internal communication.

  - **Hidden layers** have at least one layer for internal communication and evaluation.

  - **Output layers** are responsible for outputting the results.

ANN performs in two ways.

- **Feedforward**: All the nodes are connected, and the flow to the finalization direction is from the input layer's node to the output layer's node. ANNs learn from mistakes and improve their accuracy over time due to the training of the models.

---

[21] Features (a.k.a. variables, or attributes) are created/calculated accordingly to the business context, and used as inputs/parameters to train ML models and execute predictions (e.g., time of usage of a specific product, volume of purchases, type of products bought or not).

- **Backpropagation**: The process is similar to the feedforward, except its direction is the opposite: from the output's layer to the input's layer, and each node's precision is checked, and its weight is recalculated for a bigger or lower value, depending on the precision results based on each node's errors.

ANNs are adaptive and enable the machine to continuously learn and improve from mistakes. As a consequence, ANNs can help speed up tasks that were performed manually (e.g., face recognition).

They are trained with massive data, which can be raw, and they identify patterns based on that. It can also be done in the most common manner—supervised, which means that the data is labeled by humans and presented to correct data. Then, the algorithm learns to apply similar behavior for analytics or to generate data.

## Deep Neural Network

A deep neural network (DNN), also known as deep learning, is a subset of machine learning that uses the same concept as ANN. The difference is that it uses multi-layered neural networks.

Basic ANNs have up to three layers, while DNN has more than three layers and is used in scenarios where the accuracy needs to be even greater than a simple ANN algorithm.

Due to its depth of layers, the deep learning algorithms require much computer power.

The training process is similar to ANNs: massive data, which can be raw, and patterns can be identified based on that. It can also be done in the most common manner—supervised, which means the data is labeled by humans and presented to correct data. Then, the algorithm learns to apply similar behavior for analytics or to generate data.

The use cases are the same as ANNs business cases, such as image recognition, voice recognition, and emerging technologies such as self-driving cars.

## Generative AI

Although it is the subset of AI closest to the AGI's definition, it is Narrow AI.

The AGI mimics human behavior but is not able to human-act. It uses machine learning models to generate new data or content, such as text, images, music, audio, and videos, that are new and unique.

In other words, this type of AI can teach the machine to go further: be creators of data instead of only users of it.

Creating a picture or video from scratch based on a given prompt or even more robust, such as improving the customer's satisfaction through an efficient chat interface, are some examples of possible usability of generative AI.

## AI Agents

The first movements related to AI started with the Alan Turin's paper published in 1950, called *Computing Machinery and Intelligence*, discussing how computers can behave like humans. Since then, many studies have evolved. In 1995, Stuart J. Russell and Peter Norvig wrote about the concept of intelligent agents in their book: Artificial Intelligence: A Modern Approach. It became especially discussed nowadays, called as AI agents due to the advancement of Artificial Intelligence techniques and the potential of it. AI agents are systems focused on automating complex tasks, and executing processes, with planning, and decision making. They can use different types of Artificial Intelligence, generative or not. For instance, they can be simple: by executing specific tasks, through the rule-based design, by applying logic systems, be compound: by orchestrating different types of ML models and APIs, or autonomous: create, manage and execute complex tasks, without human intervention. Pragmatically speaking, their process of work is: once defined the goal of the agent, it'll plan what must be applied and split it into tasks, which will be executed using AI (generative or not). Of course that this flow will be fully aligned with the goal of it. And yes, it is also a product of data.

We've now firmly entered the AI age, and it's important to remember that AI systems are as good as the data they learn from. In other words, whether the data used is inaccurate or biased in some aspect, the performance of the AI systems is directly affected due to the potential incorrect patterns learned. Thus, it is critical to have responsible professionals for the data strategy to design and implement data as a product in adherence to business needs.

Finally, it's important to remember that every analytics, whether advanced or not, fits in the data as a product type and should be managed accordingly.

# Project Management

Project management is a methodology created to successfully plan and execute projects.

The main goal of project management is to ensure that the customers' expectations are minimally achieved.

## Waterfall

Waterfall is one technique to manage projects in that way because its structure enables a sequence of steps to be followed, considering the estimated effort.

The planning is created before the project's kickoff, with detailed steps for all activities that are implemented until the project's finalization. In other words, you create the plan at the beginning of the project after a detailed study of steps and expected milestones. Once planned, you and your team can work on the project by following the plan created and updating its status.

The graph visualization of this planning is called a Gantt chart. Henry Gantt created it during World War I. In the 1980s, computers started to become popular, and the Gantt graph was created for virtual visualization and to be used to manage projects.

## The Frameworks

There are different frameworks created by the market based on the waterfall, and new ones are usually emerging. As our objective is to share fundamentals, this book covers the concept of the most common.

### PMBOK

The Project Management Body of Knowledge (PMBOK) is a framework developed by the Project Management Institute (PMI), composed of five processes (initiating, planning, executing, monitoring and controlling, and closing) to standardize the project management steps. Although it supports improving many aspects, such as communication and standardizing the projects, it's not the most popular among the technical team because it is grounded in detailed planning and documentation.

Table 1-8 summarizes the roles and responsibilities of the waterfall approach.

***Table 1-8.*** *Project Management: Waterfall Roles and Responsibilities*

| ROLE | RESPONSIBILITIES |
|------|------------------|
| Project Management Officer | • Setting up the tools and standards for the program<br>• Define and maintain project management standards within the organization<br>• Provide support and guidance to project and program managers<br>• Manage the project portfolio, ensuring alignment with organizational strategy<br>• Financial planning and tracking |
| Project Manager | • Plan, execute, and conclude the projects according to the scope, budget, and deadlines<br>• Manage the project team and allocated resources<br>• Monitor and report project progress to stakeholders and sponsors |
| Technical Leader | • Manage the professionals, the agendas, and the internal communication among the team<br>• Monitor project progress and identify potential risks<br>• Support and collaborate with project managers to plan and execute the project activities |

*(continued)*

***Table 1-8.***  (*continued*)

| ROLE | RESPONSIBILITIES |
| --- | --- |
| Program Manager | • Plan, define, and execute the overall program and governance |
| | • Manage the main program documentation |
| | • Ensure program projects are fully done: scope implemented, quality of solutions, on time and with aligned costs planned (budget) |
| | • Manage and utilize resources across projects related to program projects |
| | A program is a group of related projects with one common strategic goal for the company. |
| Sponsor | • Ensure the project is properly funded |
| | • Lead the project by communicating with the high-level management about the project to ensure engagement and formal authorization |
| | • Provide a significant role in the development of the initial scope and charter |
| | • Serve as an escalation path |
| | • Authorize changes in scope |
| | • Provide go/no-go decisions |
| Project Team | • Execute the hands-on work for the project |
| | • Report the progress of the tasks. |

(*continued*)

*Table 1-8.* (*continued*)

| ROLE | RESPONSIBILITIES |
|------|------------------|
| Stakeholder | • Define and share the business goals and needs to compose the scope of the project<br>• Share feedback and do evaluations<br>• Actively participate in the allocation of resources<br>• Share potential risks for the business and the project during the whole journey, but especially at the beginning<br>• Advocate for the project |

# Agile

Agile methodology refers to conducting a project based on the values and principles of the Agile Manifesto. It is a set of techniques and good practices to ensure greater agility and flexibility in the project's stages, mainly for any changes that may occur along the project.

The agile manifesto was created in February of 2001 by 17 project managers after discussing all the frustrations they had and faced regarding the traditional project management approaches. Together with it, they also defined the principles of software development.

In agile, processes occur dynamically: to promote continuous improvement, the development is incremental. The deliverables are quick, guaranteeing quality and continuous feedback between customers and the team.

The following summarizes its manifesto.

- Individuals and interactions over processes and tools

- Working software over comprehensive documentation

- Customer collaboration over contract negotiation

- Responding to change by following a plan

The following defines its principles.

- **Customer satisfaction**: Our highest priority is to satisfy the customer through early and continuous delivery of valuable software.

- **Welcome change**: Welcome changing requirements, even late in development. Agile processes harness change for the customer's competitive advantage.

- **Continuous deliverables**: Deliver working software frequently, from a couple of weeks to a few months, with a preference for the shorter timescale.

- **Team working**: Business people and developers must work together daily throughout the project.

- **Motivated team**: Build projects around motivated individuals. Give them the environment and support they need, and trust them to get the job done.

- **Face-to-face**: The most efficient and effective method of conveying information to and within a development team is face-to-face conversation.

- **Software working**: Working software is the primary measure of progress.

- **Constant pace**: Agile processes promote sustainable development. The sponsors, developers, and users should be able to maintain a constant pace indefinitely.

- **Good design**: Continuous attention to technical excellence and good design enhances agility.

- **Simplicity**: The art of maximizing the amount of work not done is essential.

- **Self-organization**: The best architectures, requirements, and designs emerge from self-organizing teams.

- **Continuous improvement**: At regular intervals, the team reflects on how to become more effective, then tunes and adjusts its behavior accordingly.

## The Frameworks

The market creates many frameworks, and new ones are usually emerging. This book covers the two most common.

- **Kanban** is an agile methodology created in the 1940s by Taiichi Ohno, chief engineer at Toyota. He had the mission of improving the efficiency of Toyota's operations and reducing waste. Then, he developed Kanban, which means *board*, and allows the monitoring of tasks and the identification of the bottlenecks in the workflow. Since then, teams from different fields of activity have adopted the method due to its ease of use and adaptability.

- **Scrum** is an agile framework that helps teams manage work through a set of values, principles, and practices to develop and evolve complex products (highly emerging or with requirements that are quickly changing). Its values encourage teams to learn from experiences, organize and adapt themselves to solve problems, and continually improve their processes.

Table 1-9 helps you learn more about the roles and responsibilities in the agile approach.

***Table 1-9.*** *Project Management: Agile's Roles and Responsibilities*

| ROLE | RESPONSIBILITIES |
|---|---|
| Product Owner | • Customer's representative in the company |
|  | • Product vision |
|  | • Competitive analysis |
|  | • Product portfolio management |
|  | • Continuous monitoring of products |
|  | • Design of the strategy, tactical, and operational plan |
|  | • Prioritization of initiatives |
|  | • Refinement of the needs |
|  | • Collection of feedback |
|  | • Product backlog management |
|  | • Write the user stories |
|  | • Close relationship with the Dev team |
|  | • Deliverables |
|  | • Project management |

*(continued)*

***Table 1-9.*** (*continued*)

| ROLE | RESPONSIBILITIES |
|------|------------------|
| Scrum Master | Support the dev team |
| | • Coach the members in self-management and cross-functionality |
| | • Create high-value Increments that meet the definition of done |
| | • Help the team to eliminate blockers |
| | • Ensure agile events take place and are positive, productive, and kept within the timeframe |
| | • Find techniques for effective product goal definition and backlog management |
| | • Provide ways for the team to understand the need for clear and concise backlog items |
| | • Establish empirical product planning for a complex environment |
| | • Facilitate stakeholder collaboration as requested or needed |
| | Support the company |
| | • Embody agility through training and coaching to the organization to ensure continuous improvement and learning |
| | • Help employees and stakeholders understand and instill an empirical approach to complex work |
| | • Remove barriers between stakeholders and scrum teams |
| Dev Team | • Create a plan for the sprint, the sprint backlog |
| | • Instill quality by adhering to a definition of done |
| | • Adapt their plan each day toward the sprint goal |
| | • Hold each other accountable as professionals |

## The Best Approach

As you can see, there are pros and cons to each approach. That being said, the following is some golden advice based on these almost 20 years of experience.

- There's no silver bullet rule!

- Pick the best approach for you: Although many affectionate people point out one or other as the best, you need to define what works better for your needs.

- Decide the solution or adapt the technique to

  - your project needs

  - the company

  - the team's work dynamic

The most important thing is to solve the problem with a good product and plan!

## Product Management

*Data as a product is about applying product thinking to how data is modeled and shared. This is not to be confused with product selling.*

—Zhamak Dehghani[22]

Data started to be valued as an asset for companies over 15 years ago. Since databases were created, regardless of the line of business, the data always had customers who used them for different purposes.

---

[22] Zhamak Dehghani (2022), *Data Mesh: Delivering Data-Driven Value at Scale*, O'Reilly Media.

The importance of data created the need to productize it to deliver more scalable value through product management.

But what exactly is product management?

The renewed and updated product management concept emerged with agile in 2001 when the manifesto was created.

Product management is the function that manages the product's life cycle, which includes identifying the issue in the market, defining and implementing the solution, and its ongoing maintenance by meeting the customers' needs and desires.

Who manages the product, and what are their responsibilities?

## Roles and Responsibilities

According to the Agile methodology, specifically created by Scrum and followed by other frameworks, there's just one role responsible for product management: The product owner, who is responsible for the strategy and tactical plan. This role also involves the project management of each phase of product design and evolution (see section 1.3).

It has many responsibilities, which may be even greater depending on the business strategy and size.

The market created new titles with a redefinition of responsibilities due to a summarization of reasons, such as the obvious volume of effort and accumulated responsibilities, organizational structures, leadership's unfamiliarity with Agile's roles definitions, and so on. Besides, the market is still adapting by creating new titles and refining the responsibilities to fit the umbrella's responsibilities for the product owner's Scrum-defined role.

For this reason, this book focuses on two titles commonly used in the market: product owner and product manager.

In 2018, Marty Cagan[23] wrote his perspective for both roles in his book: product managers perform more strategic responsibilities, while the product owners were redefined as responsible for more tactical ones.

Although with redefined and split responsibilities, he reinforced that the roles are complementary and, ideally, product managers should be able to fully perform both responsibilities.

In short, since the product manager role name was created, the big scope defined by Scrum in the past for a product owner, who, as its name says, the owner of the product (from strategical to the operational scope) had changes in the market: it was split.

The product owner title got reduced scope by focusing on more tactical responsibilities, while the product manager got the scope of more strategic responsibilities.

To learn about what the market is practicing in terms of the split of these responsibilities, see Table 1-10.

***Table 1-10.*** *Product Manager vs. Product Owner*

| PRODUCT MANAGER | PRODUCT OWNER |
|---|---|
| Strategic focus | Tactful focus |
| Product vision | Product backlog management |
| Competitive analysis | Deliverables |
| Product portfolio management | Close relationship with the Dev team |
| Continuous monitoring of products | Prioritization of initiatives |
| Prioritization of initiatives | Refinement of the needs |
| Refinement of needs | Collection of feedback |
| Collection of feedback | Write the (user) stories |

---

[23] Marty Cagan (2018), *Inspired*, Wiley Publishing, Inc.

The following are characteristics needed to manage a product (regardless of the type of product).

- Deep knowledge of customer

- Deep knowledge of the data

- Deep knowledge of your business

- Deep knowledge of your market and industry

- Smartness, creativity, and persistence

# Some Frameworks to Manage a Product

As with everything in the world, there are many ways to perform responsibilities. I prefer using frameworks or creating them as much as possible to simplify the recurrent processes with best practices. The good news is that there are many professionals with the same preferences and interest in doing it, and we can inspire ourselves to create new ones, adapt to our needs, or simply use what they propose.

The market creates many frameworks, and new ones are frequently emerging. Some I created and I will share in this book, by the ways. As the goal of this book is to share fundamental concepts, you'll see some to complete the fundamental concepts. But again, if you'd like to be a good data product manager or, more than that, a good professional, keep studying!

## Designing Thinking

According to Tim Brown (2010),[24] design thinking is a mental model abstraction used to create innovative solutions by merging logical thinking with intuition and inspiration.

---

[24] Tim Brown (2010), *Designing Thinking*, Alta Books.

Design thinking consists of four main steps.

- **Immersion**: Understand the customer's problems.

- **Idealization**: Define the scope of the solution.

- **Prototyping**: Create a prototype to validate it with the user. If necessary, return to the previous steps until it's minimally refined.

- **Implementing**: Implement the final version and plan and execute its release.

## OKRs

Objectives and key results (OKRs) is a framework for strategic management based on defined objectives with clear expected results. Andrew Grove, ex-CEO and current board member of Intel created it. However, the framework became famous because of Google, which uses the management tool.

To apply the framework, two questions must be answered.

- **Objectives**: What would we like to achieve?

- **Key results**: How should we measure the results to inspect if we're on the path to achieving the goal?

As soon as both are defined, the initiatives to achieve them will be defined.

Therefore, once selected, this framework can and should be used by all levels and teams in the company to ensure everybody is working to achieve the common goal.

This framework is a good tool for defining the product strategy by keeping the whole team engaged in the common goal while focusing on delivering results.

To be efficient, the company should use it as a whole by following its mission/vision with the strategic company's OKRs in a cascaded manner.

# Backlog Prioritization

## RUT Scoring

RUT (relevance, urgency, tendency) is a tool to support the prioritization of initiatives or stories. This technique was developed by Andressa Chiara in 2016 as an adaptation of another prioritization tool called GUT (gravity, urgency, tendency), created in the 1980s by Kepner and Tregoe.

The applicability is composed of listing all initiatives or stories and evaluating together with the customers/stakeholders the weights for each one of the initiatives. The weights go from one to five scores and must be done for the three perspectives of RUT by answering the questions.

- **Relevance**: What's the relevance of the product?

  - Nice to have, but we're fine without it

  - Desired and helpful

  - Important

  - Very important

  - Critical and would deliver much value

- **Urgency**: What's the timing for the business?

  - It can wait.

  - It can wait, but it is not good for business.

  - It's important to be the next priority.

  - It's urgent.

  - It's critical. It needs to be done now.

- **Tendency**: What's the user's perspective?

  - Not good, but it can be used

  - It may be problematic to remain like that

- Problems coming soon

- The problem is growing

- Much worse and more exhausting

As soon as the weights are applied for each initiative, it's time to apply the formula, which creates the prioritization score.

- RUT score = *Relevance * Urgency * Timely*

A descending order of the RUT score value suggests the priority of initiatives/stories.

## RICE Scoring

RICE (reach, impact, confidence, effort) is a tool to support prioritizing initiatives or stories, created by Sean McBride for Intercom.

The applicability is composed of listing all initiatives or stories and evaluating together with the customers/stakeholders the weights for each one of the initiatives. The weights result from an equation created by using the four perspectives of RICE, which must be answered individually.

- **Reach**: (based on a specific period) How many users should this feature/initiative/story reach?

- **Impact**: How much will this feature/initiative/story impact the users?

  - 3x = Massive impact

  - 2x = High Impact

  - 1x = Medium Impact

  - 0.5x = Low impact

  - 0.25x = Minimal impact

- **Confidence**: What's the confidence for the estimations?

    - 100% = High Confidence

    - 80% = Medium Confidence

    - 50% = Low Confidence

- **Effort**: (based on a specific measure of time defined)
  How much effort (by a professional)?

As soon as the weights are applied for each initiative, it's time to apply the formula, which creates the prioritization score.

- RICE score = *(Reach * Impact * Confidence) / Effort*

A descending order of the RICE score value suggests the priority of initiatives/stories.

## Analytics: Measurement

OKRs include creating KPIs to measure the results. The measurement of all the work we do is critical. For this reason, I greatly enjoy the DMAIC (defining, measuring, analyzing, improving, controlling) concept, derived from Lean Six Sigma, which focuses on the following.

- **Defining** the problem to be addressed and expected outputs for them.

- **Measuring** the performance (ideally) with quantitative results to map the scenarios about processes to identify issues and opportunities actively and proactively.

- **Analyzing** the process flow by comparing the past and current scenarios to validate the correct path and whether adaptations are necessary. And simulate future scenarios for identified opportunities.

71

- **Improving** the processes by testing and implementing solutions.

- **Controlling** the implementations by maintaining the monitoring process to keep the benefits evaluated during the journey and stay on the cycle of continuous improvements.

## The Best Approaches and Fit for Data Product Management

First, there's no silver bullet. What is the ideal and best practices path here is the one designed to achieve the vision of the company. Therefore, the weight of the impact of it depends on the strategy vision of the company's data leadership.

It's also important to remember that putting the ideal and best practices in place is not an easy and smooth path. Actually, it's the opposite. It requires much work from the leader, engaging their teams, peers, and customers, considering that it'll be necessary: strong strategic vision, strong tactical planning, and work until the operation phase is performing as expected. Which, hopefully, will keep running if everything goes and stays on a successful path.

Why isn't there one best approach path to follow? Besides the obvious internal and external complexities faced in our routine, there are still the individual characteristics of the business, the market, the company, its culture, and others. Thus, there's not necessarily room (as we'd like) to implement the best practices from books or any other source of knowledge used. In my opinion, this has a good side as well because these challenges usually create lots of new opportunities and paths for best practices. I reinforce the following: except for the regulatory and legal definitions, which are hard and standardized requirements, everything else is related to the good strategy defined by a good leader, including, by the way, how to operationalize the mandatory work, as regulatory and legal requests.

One of the key abilities for leaders is to achieve and deliver great results with the best usage of the tools available.

The ideal and best practices context are not just related to how we do it, by defining the strategy and best tools for that, but are also related to who does it.

That being said, it's pretty clear the reason for the continuous evolution of the market has been creating new profiles and roles to better fit all product management responsibilities: the best organization possible by achieving results.

For these same reasons, when talking about the best-fit profile for data product management, I think we shouldn't have one specific role. It should be part of the data management role.

It's completely obsolete to expect and request that a data manager or data executive should be exclusively technical. They must work linked strategically to the business needs by applying product management to their routine (see Figure 1-4).

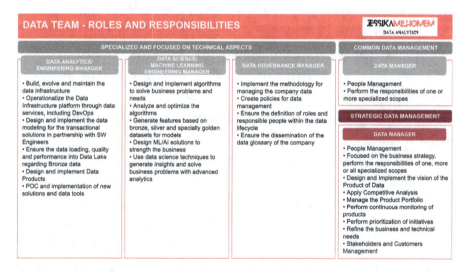

***Figure 1-4.*** *Data manager scope*

We should appropriately leverage career paths (management and specialization paths) by counting on specialized technical leaders (performed by individual contributors) to handle and operationalize super detailed technical solutions.

They are the appropriate professionals to focus on these details. Based on that, they will also be able to develop their career by growing their knowledge, experience, and, consequently, their scope.

Data managers should focus on the business strategy by leveraging, designing, and implementing data technologies and products through their teams. Data managers must perform their responsibilities strategically. After all, how do you deliver results for the business, the main customer of any data product, if you focus only on technical aspects? Everything should be part of the strategy for this role.

Clearly, its scope and responsibilities must be equivalent to the data manager's seniority level. Of course, I consider seniority equivalent to knowledge, experience, and impact generated for the business. Not just the title. It is not uncommon, but it is not the focus of this book.

Additionally, data knowledge is necessary as a specialty to manage data products. After all, the problems, needs, and evolution are fundamentally related. Even though it is at the same level of importance, it is necessary business knowledge.

In short, data managers should perform data product management responsibilities.

This is the ideal role and scope for a data manager/executive. That's what the market is missing.

Why? There are many reasons to talk about, but one of the most important (also fundamental reason for this book) is that data strategy is usually ignored or not appropriately considered by the senior leadership/executives. It needs to change by having data senior leaders/executives in the table, with minimum and good enough knowledge to discuss and plan them with other leaders. For data analytics culture and strategy, the bottom-up is not enough; the top-down vision is fundamental to guiding the company.

A clear example of this opportunity in the market is that, commonly, the business analytics needs are the focus, while the foundation for the data used for them is usually underestimated.

Data and analytics strategies should be developed together because they depend on and evolve. At least, they should. If not, at some point, the technical debts become big blockers or headaches for the business.

Table 1-11 contrasts the advantages and disadvantages of data managers performing data product management.

***Table 1-11.*** *Evaluation of Data Managers Performing Data Product Management*

| GENERAL ADVANTAGES | GENERAL DISADVANTAGES |
| --- | --- |
| Problem-solving and Productivity | Problem-solving and Productivity |
| • Technical knowledge to understand the functionalities/features that are necessary and would be helpful for technical and non-technical customers | • Learning curve: Initially, managers less experienced with product management skills |
| | Leadership |
| • Removed the detailed dependencies of product teams from data leaders regarding data concepts and applicability | • Learning curve: Initially, managers are less experienced with product management responsibilities |
| • Easier interface with customers | Culture |
| • Easier interface between the technical and business teams | • Learning curve: Necessary adaptations on the processes and expectations for data managers and product managers that worked with data |
| • All data management requirements considered in the strategy: mandatory, necessary, and desired features and solutions | |

*(continued)*

*Table 1-11.* (*continued*)

| GENERAL ADVANTAGES | GENERAL DISADVANTAGES |
|---|---|
| Leadership | • Learning curve: Necessary adaptations on the processes and expectations for data managers and data specialists |
| • Resilient leadership: Leaders with technical, business, and product management abilities to create more impact for the business in accordance with their scope | |
| • Efficient leadership: leaders that transition between strategic and operational | Career development |
| Culture | • Career development paths: necessary adjustments on the role and expectations for data managers and individual contributors of data |
| • Clear processes and expectations of data management | |
| • Full data and analytics culture influenced and supported the institutions | • Development of skills: Data managers who perform strictly technical work would need to develop skills regarding strategic and tactical vision for the product to perform data product management |
| Career development | |
| • Clear role, path, and expectations for data management (including data product management) | |
| • Clear role, path, and expectations for data specialists | • Development skills: Product Managers need to develop strategic and tactical technical skills to become data managers |
| • Managers are able to evolve in their career from the initial level (more operational and technical focused) to the executive level (more strategically business focused) | |
| • Retention of professionals due to more opportunities and challenges, enabling long career development inside the companies | |

Although this approach has its disadvantages, like everything in life, they are lower and lighter when compared with its benefits.

Of course, it's not possible to define that all data managers succeed, as it's not possible to do for any professional. It'll depend on their experience, skills, and how they apply them.

And here is the key point: the role and expectations of data managers must change and evolve! They must be focused not just on the technical aspects but also be part of the business strategy and execution, creating an impact for the business through data and technologies. The leadership must be based on data management and data product management to solve real problems!

That's why the potential for this approach is higher than any other profile once the requirements are checked.

To conclude, ask yourself some questions. As a consumer or producer of data, would you be able to clearly align your needs and problems by discussing them with a product manager? Do you feel confident with these alignments for the mid to long term when discussing them with a product manager? Would it be necessary to involve the data manager to align your issues, needs, concerns, and short/mid/long-term solutions? Do you have confidence in resolving your problems in the mid and long term after facing this experience?

The data world is different from the traditional technical products on the market. The knowledge and needs are different because the context is different. Consequently, the approach must be appropriate to it!

That said, please keep in mind that I'm considering "both roles as one" as I go through the book.

# 1.4. Summary

Let's recap the main takeaways of this chapter.

- **Problems are the source of everything**: Before designing a solution, it's necessary to have clarity about the problem, even when it's unclear. Products exist to address them. Thus, focus on understanding the problems before everything.

- **Data management and data governance**: Data management is a broader scope that encompasses everything related to data. Data governance is part of it, defining standards and best practices for data management. These concepts are fundamental to working with data, regardless of specialty. Even more so when talking about products in this area.

- **Artificial intelligence**: AI is a multifaceted discipline within the larger field of computer science that focuses on reproducing similar human competencies, such as creativity, learning from experiences, interpreting data, and making decisions equivalent to a human being in a machine. It is divided into three main types: narrow AI, general AI, and superintelligence. We're in the narrow artificial intelligence era, which has three types of machine learning: supervised, unsupervised, and reinforced.

  - **Data product**: It is a solution that boosts data management, develops data as a product, or fundamentally uses data for a specific analytical or business goal.

- **Data as a product**: Data is a product by itself due to its core capacity to solve issues through its fundamental usability for analytical, business, or decision-making purposes. It has customers (usually, but not limited to, data analysts, data scientists, and others). Its main goal is to make the data usability simple and effective.

- **Data product management role**: This requires deeper data, product, and project management knowledge. With equivalent importance, the business context and knowledge are fundamental. Strategic and leadership skills are also critical. Due to these needs, data managers are the best professionals to perform data product management responsibilities. Although success is not guaranteed just by having this professional title (as any professional title does not guarantee it), this approach has great potential. I recommend it due to the expertise of the professional and all the benefits of this approach. To finalize reinforcing here, the most important is that the role and expectations of data managers must change and evolve! They must be focused not just on the technical aspects but also be part of and lead business strategy and execution by creating impact for the business through data and technologies. The leadership must be based on data management and data product management to solve real problems!

- **Data project management**: To implement and maintain any product, it's fundamental to design and plan and manage projects. It can be done through different frameworks (based on the waterfall or agile approaches). It must be picked based on the best fit for the business needs, team context, and project needs.

# Summary

In this chapter I introduced important data product management concepts. However, you still need to understand/review the traditional architecture, which brings us to the next chapter, which discusses historical context, problems, limitations, and solutions.

It's important to have clarity about these concepts and information to understand potential problems and challenges to understand what we had in the past, what we still have and possibly infer what we'll have as problems and challenges. Besides, this knowledge is super relevant as the foundation for Part II of this book.

# CHAPTER 2

# Traditional Analytical Data Architecture

This chapter reviews traditional architectural concepts and definitions. It covers problems faced, the history and evolution of traditional architecture, and each component, such as the data warehouse, data mart, stage, and operational data store. Additionally, I explain the processes used for data wrangling and how they work.

## 2.1. Data Architecture

Why should you learn about traditional architecture? After all, we're in the AI era. Now, we work with much bigger volumes of data (not all companies, but most of them). Thus, we should focus only on that, right? No!

The rationale is old: We need to learn from history to not perform the same errors of the past. And even more, to better understand the whys, and based on the learnings, "predict" what will come.

It is possible; again, I expect leaders to know it. As a more senior professional, you're not as close to the coding and operations because you need to focus on strategy and other important aspects important for the business and to support your team. To be qualified for this responsibility,

© Jessika Milhomem 2025
J. Milhomem, *Data Product Management in the AI Age*,
https://doi.org/10.1007/979-8-8688-1315-3_2

you need to have more than a minimum knowledge of it. What about
the basic concepts? A good leader should know them and be continually
updated about the market!

Unfortunately, the market is not a Pollyanna world, and many less-
than-qualified professionals end up in leadership positions. It occurs for a
variety of reasons that are irrelevant to this book.

The truth is that, in addition to other reasons, a leader's lack of
fundamental knowledge causes a great deal of annoyance, inconsistencies,
inefficiencies, and other damages at all levels of business.

Again, I'm not saying that everybody should be a superhero. But if
you want to work in a specialized area, you need to at least study and
learn as much as possible about it to communicate and lead successfully.
Especially as a senior leader!

As a leader, I incentivize and mentor my team as much as possible to
keep learning and growing as individuals and as part of a group. Of course,
I also do it to myself because, as you may have imagined, it's one of my
core values. And it's important to be humble regarding knowledge and
know that you always have opportunities to learn.

Thus, whether the reasons for learning new technologies, the whole
context, and basic fundamentals are not enough for you yet, let's practice a
reflexive exercise by answering the following questions.

- Would you be comfortable being led by someone who
  doesn't have basic knowledge to define a strategy
  for your career? Would you be comfortable being
  evaluated by this person?

- Now, imagine that you are the CEO of the company.
  Would you be comfortable giving the same person the
  mission of heading strategy?

You'd probably answer no to both questions, right? So, why would you
want to be this uninformed person?

# The History of Data Architecture

Before explaining each one of the pieces of analytical architecture, let's look at the timeline of traditional data architecture.

> **1950s and earlier**: Punch cards and magnetic tape were used for data storage. Punch cards (a.k.a. the IBM card) became a symbol of 20th-century automation that came with a catchphrase warning: "Do not fold, spindle, or mutilate." Magnetic tape, inspired by vacuum cleaners,[1] increased the speed of data processing and removed the dependency on massive stacks of punched cards for data storage. Also very important, Alan Turin wrote the paper Computing Machinery and Intelligence, discussing how computers could behave like humans - the start of AI studies and evolution.
>
> **1956**: IBM was primarily responsible for the early evolution of disk storage with the creation of the first random-access disk drive, called IBM 305 RAMAC.
>
> **1970s**: Edgar F. Codd, a mathematician who worked at IBM, published an article called "A relational model of data for large shared data banks."[2] It supported not just IBM but the market as a whole in understanding how to directly access data through software, which created different commercial relational database management systems, such as the first one: Oracle.

---

[1] IBM, *Magnetic tape*. Available at https://www.ibm.com/history/magnetic-tape

[2] E. F. Codd, IBM Research Laboratory (1970), *A Relational Model of Data for Large Shared Data Banks*. Available at https://www.seas.upenn.edu/~zives/03f/cis550/codd.pdf

**1980s:** Relational databases became popular among businesses, and many companies assigned personal computers to their staff.

**1988:** Robert Shaw published *Database Marketing* to introduce data usage for business analytics needs by suggesting merging marketing principles and computing techniques into a database marketing program to increase sales.

**1990s:** The internet was emerging in popularity, and the competition had increased, requesting a structured way to work with strategy: by using business intelligence.

**1991:** William Inmon published *Building the Data Warehouse* by creating the inspirational data warehouse architecture.

**1998:** Ralph Kimball and Margy Ross published the *Data Warehouse Lifecycle Toolkit,* which introduced innovative perspectives on data modeling techniques for a data warehouse. These concepts are still referenced today.

# Problems

Evolution is created upon problems and challenges faced, and this was not different for the data analytics journey. Actually, it is the key to understanding how the current data analytics architecture has emerged and why it is still important, even more so in the era of AI.

To better understand how the traditional analytics architecture was technically designed, let's first talk about the problems faced at that time (1980s–1990s), even though many are still present.

- **Strong competition in the market due to globalization:** This included profound transformations in the international economic system and organization of work. Now, the concurrence was domestic and international, with the proliferation of transactional companies.

- **The necessity of data for analytics:** Due to the challenges in the market, new disruptive and innovative alternatives were necessary for the emerging new world, and data analytics was the big bet for differentiating companies. Those who can better evaluate their data have more opportunities to become more efficient. Nonetheless, the use of their own data also had lots of challenges.

- **Many systems as sources for analytics:** With the invention of relational databases, many operational systems were created to manage specific operations. The large volume of existing data sources created much effort for the employees who needed the data to be able to analyze them.

- **Full data for analytics:** Having only pieces or windows of data available was not enough to support analytics. For a better understanding, the data needed to be related to the present (most recent information as possible) but also enable the historical data.

- **The necessity of a common data model:** A large volume of data sources also created difficulty in understanding it and making better decisions because, as they were evaluated in silos, it was common to have redundancy, inconsistency, and proliferation of decision support systems.

85

- **Not impacting the transactions happening on the operational database**: The operational databases were developed and customized to support transactions and performed well when operating with ACID (atomicity,[3] consistency,[4] isolation,[5] durability[6]). Thus, querying massive data in the operational system was not an efficient and secure option, as it could hugely impact the business operation.

# Enterprise Data Warehousing

New ideas using data for analytics were already emerging due to the challenges faced in the market, such as database marketing, which was suggested by Robert Shaw in 1988 to focus on marketing needs.

In 1991, Bill Inmon wrote *Building the Data Warehouse*, describing a data warehouse architecture with detailed technical guidelines. The innovative ideas were a great fit for the sedent of solutions from the market. It was the beginning of the analytical architecture design. Besides Inmon, "the father of data warehousing," Ralph Kimball also wrote a book introducing innovative perspectives on data modeling techniques for a data warehouse. We're going to understand both perspectives because they are still references for the current day.

Curiously, many people who work with data don't know much information about it.

---

[3] Atomicity: The entire transactional takes place at once or doesn't happen at all.

[4] Consistency: The database must be consistent before and after the transaction.

[5] Isolation: Multiple transactions execute in an independent manner and not impact each other.

[6] Durability: A successful transaction is committed and persisted in the database, even if there are system failures.

# What Is a Data Warehouse?

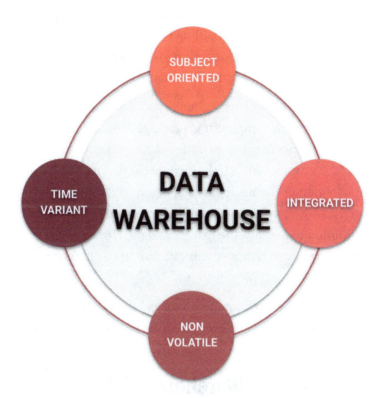

***Figure 2-1.*** *Enterprise data warehouse definitions*

A data warehouse (a.k.a. an *enterprise data warehouse*) is the central and specific repository for analytics. It can be defined as an ecosystem for the traditional environment (in the past, a.k.a. a *corporate information factory*) or a specific piece of this architecture, depending on the author of it.

The universal definition is that it is an architecture that supports business decisions and is based on four main characteristics defined by Bill Inmon, which form the foundation for the analytics environment (see Figure 2-1).

- **Subject-oriented**: It holds all business domains.

- **Integrated**: Integrates all data sources by consolidating the full information, enabling insight into operational processes and new possibilities for leveraging analytics to make decisions and create business value.

- **Non-volatile**: It doesn't change its history. With Inmon's approach, it's never changed: the data is just inserted, never updated or deleted, to enable auditing of the output. With Kimball's approach, once the data is loaded, it can be updated at maximum for specific cases when a slowly changing dimension (SCD) is applied to manage the historical data. However, the data has been removed from the database. It's critical to keep the complete history.

- **Time-variant**: It contains historical data, changes over time, and the most current data available for analysis.

# What Does a Data Warehouse Solve?

The architecture of a data warehouse, especially based on the corporate information factory, can solve the following issues.

- The necessity of data for analytics

- Many systems as sources for analytics

- The necessity of full data for analytics

- The necessity of a common data model

- Does not impact transactions happening on an operational database

The solution is applied to them with its design capabilities.

- Isolated architecture because it's apart from the operational environment

- Scalable through modular and integrated architecture

- Different granularities of data containing operational and consolidated data

- Accurate data treated with business rules to support business decisions

# Inmon's Vision

For Inmon, the data warehouse is the main part of business intelligence (BI) architecture. It is the heart of the architecture and the foundation of all decision support systems processing.

*Building the Data Warehouse*[7] provides a good visualization of architectural design. However, the following summarizes Inmon's vision.

- **Single and granular**: This is a single integrated source of data to perform analytics, with granular data with easy accessibility for reusability and conciliation. The data warehouse has different levels of granularity by layer.

  - **Older level of detail** (a.k.a. cold data): Alternate bulk storage and storage of all the historical data, especially for security and regulatory purposes.

---

[7] William H. Inmon (2005), *Building the Data Warehouse* (Figure 2-5: The structure of the data warehouse, p. 34), Wiley Publishing, Inc.

- **Current level of detail** (a.k.a. hot data): The most recent information from one or two years back in history is continuously updated with operational data sources. As the data ages, it's moved to the "older level of detail" layer.

- **Data mart**: Lightly summarized data is updated from the "current level of detail" layer consolidated with stored data and eventual additional predictions. This layer can store historical data due to consolidation's lower volume of data.

- **Highly summarized**: Highly summarized data is updated from the "data mart" layer and consolidated with its own stored data. This layer can store all historical data due to the lowest volume of data compared with all layers, resulting from high aggregation.

- **Subject-oriented**: A data warehouse is organized around the company's functional applications. In other words, by business domains. For instance, for a pharmaceutical company, the major subject areas might be medication, diseases, customers, orders, vendors, bills of material, and raw goods.

- **Integrated**: A data warehouse is integrated and fed with data from all available and necessary data sources. Once integrated, the data is treated and transformed into one unique and standard output. The encoding is consistently applied regardless of method or source application.

- **Non-volatile**: It doesn't change its history because once the data is loaded, it'll have snapshots of data, and nothing can be removed from the database to keep the complete history. For example, the snapshot data is loaded into a data warehouse. Thus, as soon as a change occurs in the operational data environment, a new snapshot is loaded into a data warehouse to represent this change.

- **Time-variant**: The historical and most current data available for analyses. The data warehouse has sophisticated snapshots of data that do not change the historical data and contains different layers in the stack to ensure the time window for historical data that stores from five to ten years of historical data. Meanwhile, operating systems update data records, don't have mandatory definitions of time, and traditionally only store one to three months of historical data.

The data modeling is done by normalizing the database in the 3NF.

# Kimbal's Vision and Definitions

For Kimbal, a data warehouse is a generic ecosystem for BI architecture that follows the bus architecture design. It's a parallel design to a corporate information factory.

*The Data Warehouse Toolkit*[8] provides a good visualization of this architectural design.

---

[8] Ralph Kimball and Margy Ross (2005), *The Data Warehouse Toolkit* (Figure 1.1: Basic elements of the data warehouse, p. 7), Wiley Computer, John Wiley & Sons, Inc.

- **Same values, but different architecture**: The core values of a data warehouse are in accordance for both authors; however, the architecture has divergences in its designs. There are different layers of data to manage the same granularity with techniques for data modeling.

  - **Granularity**: Different from the enterprise data warehouse described by Inmon, the granularity is managed by each subject (nowadays called data domain), considering its business and particularity needs. Kimball reinforces the modeling techniques (based on conformed dimensions and facts) that enable the users to quickly answer their questions at any level of analytics. Besides, historical information is also treated with data modeling techniques, such as SCD.

  - **Data staging area**: A transient area that stores operational data sources and where the ETL process is executed to load the final data into the data warehouse.

  - **Data presentation area**: This is where the accurate and available data is. It is accessible by users through queries and presentation tools. Kimbal refers to the union of all data marts as the data warehouse, as they together represent all business subjects. The subject creates the data marts and has the necessary granularity due to dimensional modeling. This technique has become popular and is still inspirational today because it brings much simplicity to discussions and understanding with business professionals, besides the direct technical results.

- **Data access tools**: This is the last layer of the architecture and is where the tools are allocated to query the data in the presentation layer.

You'll see it in the next chapter, but it's worth saying that the logical visual of Kimbal's architectures is a clear inspiration for the design of big data architectures. They have similar layers, also considering solutions designed by Inmon.

# Staging Area

The staging area is commonly a transient layer of data that works as a storage for the extracted operational data (a.k.a. raw data) and for the ETL (extract-transform-load) process until the complete loading of data into the data warehouse.

This layer is part of the BI architecture for both approaches.

According to Inmon, the staging area is especially important for larger amounts of data with high complexity of processing, as well as for coordinating data integration from different sources and availability.

These two last characteristics of data sources are important because the synchrony of the raw data is necessary to make sure the data is accurate and compliant with the four characteristics (which I also consider as values) of the data warehouse: subject-oriented, integrated, non-volatile, time-variant.

Kimball[9] reinforces that the staging process is a critical component of the data warehouse because all data processing is there to ensure the data is accurate. This is even evidenced in their data warehouse architectural proposal.

---

[9] Ralph Kimball and Margy Ross (2002), *The Data Warehouse Toolkit: The complete Guide to Dimensional Modeling*, John Wiley & Sons, Inc.

However, being critical doesn't mean that it should be accessible to anyone. That's why the staging area is unavailable for users through queries or presentation services.

# Data Marts

A data mart is subject-oriented (a.k.a. domain-oriented). It is often defined by the department. It enables data users to answer their most common questions and generate new insights.

The data source for a data mart can originate from the data warehouse or the staging area, depending on the author of it.

## Inmon's Definitions

Inmon classifies the data mart as one of the four levels of a larger architecture (a.k.a. corporate information factory).[10] There following describes the four levels.

- **Operational** contains the operational data (a.k.a. raw data) from the applications. They are detailed, application-oriented, and hugely accessed for daily usability, and the current data is the most valuable. It answers the following type of question: What is the current credit score of this customer? It is used for high-performance transaction processing.

---

[10] William H. Inmon (2005), *Building the Data Warehouse* (Figure 2-5: The structure of the data warehouse, p. 34), Wiley Publishing, Inc.

94

- **Atomic/data warehouse** contains the most granular data integrated among all operational data, with current and historical data that is never updated and subject-oriented. Users use this level to discover data. It answers the following type of question: What is the historical credit score of this customer?

- **Departmental/data mart** contains just its subject-oriented (usually departmental) and is designed based on end-user requirements. It answers the following type of question: How many customers have we onboarded?

- **Individual** is composed by the user with individual heuristics analysis created during ad hoc analysis. It answers the following type of question: What are the patterns for this group of customers?

A data mart also has the following characteristics.

- **Data marts and data warehouses are different components**: Each component has its specific architectures, as described by the level of architecture of a corporate information factory.

- **Implemented through different technology**: A data mart is often implemented with multi-relational databases and tools that are different from a data warehouse.

- **Summarized and aggregated data**: Another characteristic is that the data marts are aggregated: lightly or highly summarized. Consequently, it has less volume of data, even having big periods of it.

# Kimbal's Vision

For Kimbal, a data mart is a process-oriented and measurement-intensive subject area that contains atomic and aggregated data. Let's evaluate each characteristic.

- **Process-oriented**: The data mart is a process-oriented and measurement-intensive subject, which means that its design is based on the business process (a.k.a. domain), considers the metrics that would be necessary and critical for that process, and not necessarily dependent on anticipated user's questions.

- **Contains atomic data**: The data marts contain atomic and aggregated data to enable analyses from different perspectives. This approach enables us to answer common business questions and make discoveries from users. It's also recommended that atomic data for modeling be prioritized for these reasons.

- **Operational data is the source**: The source for the data mart is the operational data, which is a replica of the source consumed directly. It's used to implement the data presentation area, where the data marts reside and are the unique source of the user's subject data. It is important to have the data clearly in mind.

- **Centralized or decentralized implemented**: The data marts can be created in phases until they compound the entire enterprise data warehouse architecture. The goal is to focus on incremental implementations rather than implementing the entire data warehouse at once.

# Operational Data Stores

The web stores two types of data: transactional data originating from operations and clickstream (a.k.a. events of a customer in a website), which are logs of customer behavior in e-commerce.

Both are important and relevant. The first one is part of the operational process.

The second one of the main reasons for the creation of the operational data store (ODS) is clickstream.

The clickstream is super valuable for analytics purposes based on massive history data because it allows users to create profiles that can be used to customize interactions with customers, such as product offerings.

Profiles are stored in the ODS with aggregated customer information and preferences.

ODS was designed and proposed by Inmon (2005) to bridge the analytics data and the operational data for online transactional processing (OLTP) needs on the web and for analytics.

*Building the Data Warehouse*[11] provides a good visualization of ODS design.

The ODS contains integrated data from a data warehouse and supports high-performance transaction processing. The data workflow follows these rules.

- A data warehouse receives data directly from the web or the granularity manager. This solution cleans the data and commonly reduces the big amounts by 90% by keeping just relevant data.

---

[11] William H. Inmon (2005), *Building the Data Warehouse* (Figure 10-3: Data is passed to the ODS before it goes to the Web, p. 292), Wiley Publishing, Inc.

- A data warehouse is responsible for storing the data and creating profiles with the most recent customer information to feed the ODS.

- The ODS can also feed a data warehouse or make analytical requests.

- The web connects directly to ODS, which is customized to answer requests quickly, as an operation environment is commonly needed.

- The web can connect to a data warehouse for direct queries in exceptional scenarios. The rationale is to use the ODS as the bridge for this communication due to its design, which integrates well with both architectures: operational and analytics.

For Kimball,[12] the ODS was rarely necessary when the book was written. The use cases were generally used to manage specific reports.

More than 20 years after these books were written, we see how this type of solution was important, as addressed in big data architecture.

## 2.2.  Data Ingestion and Treatment

ETL (extract-transform-load) is the process of pulling and ingesting data from the operational system (as is), transforming it into corporate data by applying business rules and standards, and then loading it into the data warehouse.

There are books focused on the ETL process. Thus, if you'd like to learn more about it, I encourage you to look further into this interesting topic.

---

[12] Ralph Kimball and Margy Ross (2002), *The Data Warehouse Toolkit: The complete Guide to Dimensional Modeling*, John Wiley & Sons, Inc.

This book focuses on the fundamentals, which is the baseline for the next steps.

Half of the heaviest intelligence of a BI solution is derived from the definition of data modeling, and the other half is in the ETL.

For the traditional architecture, there are many solutions with interfaces to automate the ETL processes. Usually, these tools are the solution to put the ETL in place due to the efficiency that they create for the work.

# Extraction

Extraction is the first step in the process of the ETL and is responsible for getting data for the BI environment, regardless of the definition of data warehouse architecture.

The process is based on the automation of reading data from the source ( usually are operational sources) and from transactional services to load them into the staging area.

The data sources can also be static files with historical data or spreadsheets for recurrent processes.

In short, they are any source of data that enriches the analytics perspective with additional information. It also includes external data, such as data bought from third parties.

# Computing

For the traditional analytical environment, only structured data was possible, and semi-structured data was starting to be accepted in the big data era.

Besides, even with components and techniques such as the ODS to work with more updated data, in traditional architecture, data computing is often done in batches.

# Scheduling Data Extraction and Incremental Data Extraction (and Loading)

Another important definition of the extraction layer is the scheduling process to ingest data from the sources. It's important because the agenda to start the data extraction and the time window selection for the delta determines how much the data warehouse is updated for any analytics needs.

There are two types of extraction (and loading).

- A **full batch** consists of reading the full data source and ingesting it (extracting and loading) from the source to the staging area.

- **Incremental data** consists of reading a piece of the data source based on a specific delta rule and ingesting it (extracting and loading) from the source to the staging area.

- **The Delta rule** can be a time window definition (such as by day, hour, or minutes) or a system's key window definition (such as filter from the maximum ID stored in the staging to the most recent ID from the operational system).

In the BI architecture, the batch process is generally implemented incrementally, with occasional full batch extraction, usually for the initial load.

The delta is usually applied by following the extraction execution, followed by the rest of the ETL process.

- **D-1** (one day before the current date): Always follow the day before to ensure the most updated information about the last day.

- **By hour**: Occasionally by hour, depending on the critical business needs, the volume of data, and the capacity of the data architecture. Remembering that the architecture uses the traditional infrastructure for it.

- **By minute**: Rarely by minutes, depending on the critical business needs, the volume of data, and the capacity of the data architecture. Remembering that the architecture uses the traditional infrastructure for it.

Observe that this technique is not new and was created for the new AI era. It's related to the limitations faced in traditional architecture that require a solution. I'll talk more about it in the "Limitations" section.

Once the data needed is available in the staging area, it's time to go further into data transformation.

# Transformation

The transformation phase is the most important step in ETL; it is where most of the work and intelligence lives. Basically, all the definitions for all components of a data warehouse, regardless of its architecture definition, exist only if the transformation step is very well designed and implemented. That's why it's where half of the heaviest intelligence of the solution lives.

Once the data is extracted to the staging area, the transformation journey begins. The following are some categories of work that are done during this phase.

## Cleansing of Data

- Correct misspellings
- Resolve domain conflicts

- Deal with missing elements
- Parse data into standard formats

# Standardization

- Naming conventions
- Logical conversion of data and datatypes
- Conversion from one database management system (DBMS) to another
- Creation of default values when needed
- Addition of time values to the data key
- Restructuring of the data key
- Deletion of extraneous or redundant data
- Ensure the correct and appropriate time horizons associated with the data

# Integration

- Data encoding
- Combine data from multiple sources
- Merge records
- Solve conflicting keys
- Deduplication of data
- Assign warehouse keys

## Business Rules Implementation

- Domain verification

- Measurement of attributes

- Summarization of data

- Apply techniques for historical data to ensure non-volatility and time-variant values

  - Implement snapshots

  - Implement SCD types

  - Automate storage for each appropriate layer

  - Automate movement of historical data to the appropriate layer

  - Ensure maintenance of historical data available in each appropriate layer

Once the data is fully transformed, it can go further in the data loading process.

# Loading

The data loading step happens at least twice. Depending on the data warehouse architecture, you define the staging area and the data warehouse.

Part of the definition of loading data is defined in the transformation step by filtering the data that must be selected and loaded after treatments.

For the traditional architecture, there are many solutions with interfaces to automate the ETL processes. Usually, these tools are the solution to put the ETL in place, including the data loading, due to the features and efficiency they create.

For massive data loading, some strategies need to be considered and planned.

- Define an efficient process to load the data and the indexes.

- Parallelize the processes for data loading.

- Use the staging area to load the data incrementally.

## 2.3.  Limitations

The main point of this chapter is to understand the issues and limitations faced in the past, learn from them, relate them to the current scenario, and predict issues.

We're facing many issues now that we've already faced in the past. We "just" need to figure out how to use and evolve the technology to solve them. Of course, it's directly related to developing data products and data as a product.

That's why I'm so incisive in saying that leaders should study, understand, and know the concepts and details, not just about current and future technologies but also their previous ones.

The point is that the issues lived in the past may be the same in the future, especially when the objectives are similar but more scalable. This is discussed more in Chapter 3.

Many business needs were addressed with the traditional analytics architecture, but many challenges were faced in implementing it. There were also some limitations, even after the solutions were designed and proposed. Technical issues, in general, were related to storage, computing, and data governance.

# Storage Limitations
## When Defining the BI Architecture

- **A large volume of data and space management issues**: Compared to the operational environment, the data warehouse has a much larger volume because it integrates data from many sources and stores long historical data for analytics. For this reason, it was fundamental (and still is, but on a different scale) to apply techniques to manage the storage to mitigate issues as much as possible.

    - **Subscribe comprehensive data storage**: As one of the business contexts, that time was the beginning of globalization. The companies started to become multinational, requesting the fundamental requirement of a unique repository for analytics with the source of truth data and supporting local specifics. Clusters become part of the infrastructure to scale the necessary architecture. Thus, huge financial investments were necessary to accommodate growing volumes of historical data without compromising performance.

    - **More efficient space**: The data is stored on different spaces of the disk, depending on the storage process. For this reason, data compression is a key part of a data warehouse. When data is compacted, it can be stored in a minimal amount of space. Whether data is stored in a small space, the access to it is more efficient, and the output of the I/O is bigger for the professionals implementing a data warehouse. This technique may generate some overhead with decompactation, but even so, it's more beneficial in a data warehouse stack, as the I/O resources are scarcer compared to CPU resources.

- **Specific modeling and storage definitions**: the whole data warehouse design, regardless of the definition selected, was designed to be as efficient for queries and historical storage as possible by having specific definitions in the data modeling and time of storage, such as the dimensional modeling, suggested by Kimball, and the aggregated and old detailed data level layers, suggested by Inmon.

- **Data retention definitions**: To manage the volume of data, data retention was also critical to creating clear policies and layers for retaining or archiving data based on its significance and utility.

## After Defining the BI Architecture

- **Inability to store unstructured or semi-structured data**: With the popularity of the internet and new products and services emerging, new types of data started to emerge. These data started to become important for businesses to enable analytics on them beyond the structured data. Forcing new evolutions in the data architecture.

- **High cost and difficult to scale**: while clusters were subscribed to scale the architecture, the volume of data kept growing exponentially, requesting powerful machine resources, which demand scale by using horizontal infrastructure (replacing servers with others with more capacity). However, two main issues are related to them: powerful computers are expensive, and even having them is not optimizable, as when in a peak, it can face limits of resources; when not, it may stay idle.

- **Back-feed between operational and analytics**: With new services emerging on the internet, the capacity of architectures that could interact with low latency with the internet efficiently and securely has become critical for the business. The ODS design concept needed to evolve to support deeper integration between the operational and analytical environment.

# Computing Processing Limitations
## When Defining the BI Architecture

- **Issues for massive data loading**: As a data warehouse integrates data from many data sources and stores long historical data for analytics, it was fundamental (and still is, but on a different scale) to apply best practices techniques to manage the data loading of massive data, to mitigate crashes, and most important: the usability of the data by business analytics processes, such as designing efficient strategies to load data and indexes, parallelizing processes for data loading, and using the staging area to incrementally load the data.

  The urgency of the data is greater, especially because many processes have started to depend on these operations.

107

- **Long time to execute the full historical data**:
  Executing the full historical data is not efficient or
  cheapest in terms of technical aspects, as every day,
  it'll request a full scan from the source to load the data,
  plus the transformation. In short, lots of resources and
  money. Besides, in terms of analytics, it's not necessary
  for the business, as one of the most important values
  of the analytics environment is to be time-variant but
  not volatile, which means that the data warehouse
  must store the historical data, but not by removing and
  replacing it every day.

  It can generate mismatch issues for the historical
  data, as we know that the operational systems suffer
  updates. If replaced in the analytics environment, it
  can change the historical results. Besides, executing
  the full historical data takes longer to have the data
  available, which is also not helpful for the business.
  Thus, incremental data loading using delta definitions
  became strongly used as a best practice for these
  reasons. Full data loadings were considered just for
  specific cases based on the business requirements.

  It is another example of again facing issues from the
  past. Of course, it's also related to the time it takes
  to create innovations, but even with the opportunity
  for it or the solution available, many professionals
  underestimate it.

- **Long time to query the data**: One very important
  requirement for users of an analytical environment is
  to receive the results of a query in a short period. It's
  important because discovery or recurrent analytics
  requires lots of interaction. Thus, a slower query

response impacts the efficiency of the professionals. It's even worse when finding the information to be evaluated is complex. To address this issue, Kimbal proposed dimensional modeling to keep a simple architecture for usability, presenting data in a short answer to queries.

- **Inability to feed operational services with analytics data**: As soon as the analytics data became part of the operation of the business, it also became necessary for the operational services, such as having customer relationship management (CRM) fed with analytics data or operations that need to have aggregated analytics data in the system. In short, the value of analytics should go beyond the internal analytics environment and be able to interface with the web to support the business further.

  To solve this problem, Inmon proposed the ODS solution as a hybrid repository that supports analytics and operational requests and serves as the bridge with the data warehouse.

## After Defining the BI Architecture

- **High cost and difficult to scale**: When clusters were subscribed to scale the architecture, the volume of data kept growing exponentially, requesting powerful machine resources to store and process the data. Initially, scalability is achieved by using horizontal infrastructure (replacing servers with others with more capacity). However, it was becoming not healthy anymore, neither technically nor financially, as two main issues are related to them: powerful computers are expensive, and even having them, it's not optimizable, as when in a peak, it can face limits of resources, when not, it may stay idle. Therefore, the difficulties keep forcing the evolution of the technology.

- **Ability to use real-time analytics to integrate with services with lower latency and large volume**: As new services emerged on the internet, it became critical for a business to have the capacity of architectures that could interact with low latency with the internet, enabling back-feed between operational and analytics, by performing efficiently and securely. Besides that, with a new world more connected to the internet, new sophisticated advanced analytics emerged, becoming desired and necessary tools for the business.

  Thus, the ability to use new ML/AI technologies requesting huge data processing to train the models with massive data becomes pretty critical. Consequently, the ODS concept needed to evolve to support deeper integration between the operational and analytical environment. And again, we see the issues forcing the evolution of technology.

# Data Governance
## When Defining the BI Architecture

- **Necessary tools and best practices to manage a data warehouse**: Just as a data warehouse can create a great positive impact on the business, it can also create a negative impact if it's poorly managed. For this reason, it's pretty important to be able to monitor the environment and the results of it as much as possible. A mechanism created at that time (and still used and necessary) was the metadata design, which is "the data about the data." With it, it was possible to address questions such as what the data model is responsible for. What are the relationships among the data? What are the operational sources or any other source for it? Where are there issues in the data? With all this information, we can manage it much more easily than without it.

- **Data quality and consistency**: Integrating data from different sources and creating a source of truth is a challenge. For that reason, the ETL concept and the data modeling best practices were created. It began defining data governance, which focused more on the technical aspects.

## After Defining the BI Architecture

- **Legal compliance**: With data becoming so critical and powerful, it also became critical to have common and clear definitions for the usability of the data and how to manage it. It's necessary not just for the companies, but also for the society. As a result, at the beginning of the era of big data, new data laws were created to monitor and audit it. The laws can differ by country, such as the GDPR, LGPD, and others. As the data keeps evolving, the laws are also up to change and evolve.

- **Data governance**: Due to the big ecosystem and the importance of data for business, implementing robust practices to ensure data quality, security, and compliance with laws and regulations became critical. Also, establishing clear policies for data storage, access, and usage of internal (such as operational data) and external data (such as social media, devices, and public information) from different sources became critical to help with standards to maintain the integrity of historical data. Sometime after the definition of BI architecture, DAMA was formed and rapidly became popular. However, as data evolves, best practices for data governance do too.

# 2.4. Summary

Let's recap the main takeaways of this chapter.

- **Business problems**: The primary business problems that incentivized the creation of the traditional analytics environment were (and still are) the necessity to address the following.

    o   Strong competition in the market due to globalization

    o   The necessity of data for analytics to be more efficient and stand out from competitors

    o   The necessity of consuming data from many systems for analytics

    o   The necessity of full data for analytics (most recent and historical data)

- ○ Use a common data model for analytics instead of many silos

- ○ Perform analytics without impacting the operational databases, responsible for business transactions

- **Data warehouse**: Although the main authors have different visions for implementing a data warehouse, the common definition is that it is subject-oriented, integrated, non-volatile, and time-variant.

  - ○ **Imnon's definition** is composed of four main layers of data.

    - ■ The **older level of detail** (cold data) is an alternate bulk storage and storage of all the historical data, especially for security and regulatory purposes.

    - ■ The **current level of detail** (hot data) contains the most recent information and is moved to the "older level of detail" layer as soon as the data ages.

    - ■ A **data mart** contains lightly summarized data, with consolidated data from the "current level of detail" and eventual additional predictions.

    - ■ **Highly summarized** contains highly summarized data coming from the "data mart" layer and can store all historical data due to the lowest volume of data.

- o **Kimball's definition**

  - ■ **Customized granularity** is managed by each subject (nowadays called data domain).

  - ■ The **data staging area** is a transient area that stores operational data sources and auxiliary datasets used during the ETL process.

  - ■ The **data presentation area** is the layer where the accurate and available data is. It is accessible by users through queries and presentation tools.

  - ■ **Data access tools** is the last layer of the architecture and where the tools are allocated to query the data in the presentation layer.

- • **Data marts**: The definitions differ according to the authors of traditional architectures.

  - o **Imnon's definition**

    - ■ Data marts and data warehouses are different components

    - ■ Often implemented through different technology

    - ■ Summarized and aggregated data

  - o **Kimball's definition**

    - ■ A data warehouse is composed of data marts.

    - ■ It's process-oriented by data domains.

- It contains atomic data.

- The operational data is the source for the data mart.

- Their implementation can be centralized or decentralized.

- **ODS** was proposed by Inmon as the solution to store clickstreams and logs of customer behavior in e-commerce. It enables the usage of data to customize interactions with customers, such as the offer of products.

- **ETL** is the main intelligence of the analytical environment kept in the extraction, transformation, and loading process. The data is extracted from the sources, transformed with business rules and data quality definitions, and loaded into the data warehouse.

- Besides the business issues, **technical challenges needed to be solved and addressed with the traditional analytical architecture**.

  - Storage

    - Large volume of data and space management issues

    - Subscribe to comprehensive data storage

    - Less and efficient space

    - Specific modeling and storage definitions

    - Data retention definitions

- o   Computing
    - ■ Issues with massive data loading
    - ■ A long time to execute the full historical data
    - ■ A long time to query the data
    - ■ Inability to feed operational services with analytics data
- o   Data governance
    - ■ Necessary tools and best practices to manage a data warehouse
    - ■ Data quality and consistency
- Although many issues were solved, **others emerged and increased in the technical aspects**.
    - o   Storage
        - ■ Inability to store unstructured data
        - ■ High cost and difficult to scale
        - ■ Inability to feed operational services with analytics data
    - o   Computing
        - ■ High cost and difficult to scale
        - ■ Inability to use real-time analytics to integrate with services in lower latency and big volume

- ○ Data governance

    - ■ Necessity of defining legal compliance rules for the market to protect sensitive information

    - ■ The necessity of the business to be compliant with legal requirements

    - ■ The necessity of data governance definitions for best practices to manage data

    - ■ The necessity of the business to be compliant with data governance requirements

## Summary

In this chapter we discussed important concepts in traditional architecture, the issues that generated it, and challenges and limitations. Next, let's talk about the next phase: big data architecture. The next chapter continues the analytics journey by talking about the continuation of its history, its problems, the solutions that emerged, its limitations and challenges, and how the problems were fully, partially solved in WIP, or not solved at all.

This knowledge is also super relevant as the foundation for the second part of this book, which covers managing data as a product in practice.

# CHAPTER 3

# Big Data Architecture

This chapter explains big data architecture concepts and definitions.

It covers the history and evolution of big data architecture and some fundamental components and concepts, including data lakes, data lakehouses, the medallion architecture, and big data architectures like Lambda and Kappa. Additionally, I'll explain the problems that inspired definitions, what they solved, and new and continuing challenges.

I don't cover technologies because that's not a focus of this book, but giving clarity about the needs, reasons and context. Then, with the knowledge, you'll be able to evaluate it by yourself, as necessary. There are many other books and materials that do, however, including full books dedicated to specific technologies and tools.

My goal in this book is to cover the most important big data architecture concepts. Although they are relevant and critical, I see many professionals who don't understand these concepts. More importantly, I want to present foundational knowledge for the Golden Data Platform, a solution for creating and managing data as a product in the era of artificial intelligence (AI). It is covered in Chapter 5.

© Jessika Milhomem 2025
J. Milhomem, *Data Product Management in the AI Age*,
https://doi.org/10.1007/979-8-8688-1315-3_3

# 3.1.  Data Architecture

## History

Before explaining the innovations related to the so-called big data ecosystem, let's look at the history of data architecture.

> **2000s**: Commercial and public internet usage grew exponentially, and new types of internet usage, such as social networking, emerged. These emerging tools started requesting evolution in the data architecture to enable storing and computing a big volume of data.

> **2001**: Shift-left is created by Larry Smith.

> **2002**: Dan Linstedt introduced the data vault modeling architecture.[1]

> **2003 to 2006**: Google published papers about the Google File System,[2] Map Reduce,[3] and BigTable,[4] the technologies used as reference to create their search index.

---

[1] Dan Linstedt (2002), "Data Vault Series 1 – Data Vault Overview." *The Administration Newsletter*. Available at https://tdan.com/data-vault-series-1-data-vault-overview/5054

[2] Google, Sanjay Ghemawat, Howard Gobioff, and Shun-Tak Leung (2003), *The Google File System*. Available at https://static.googleusercontent.com/media/research.google.com/en//archive/gfs-sosp2003.pdf

[3] Google, Jeffrey Dean and Sanjay Ghemawat (2004), *MapReduce: Simplified Data Processing on Large Clusters*. Available at https://static.googleusercontent.com/media/research.google.com/en//archive/mapreduce-osdi04.pdf

[4] Google, Fay Chang, Jeffrey Dean, Sanjay Ghemawat, Wilson C. Hsieh, Deborah A. Wallach Mike Burrows, Tushar Chandra, Andrew Fikes, Robert E. Gruber (2006), *Bigtable: A Distributed Storage System for Structured Data*. Available at https://static.googleusercontent.com/media/research.google.com/en//archive/bigtable-osdi06.pdf

**2006 to 2008:** Yahoo created the Hadoop framework, inspired by Google's papers, and donated it to the Apache community.

**2008:** Inmon introduced DW 2.0, where the data life cycle is reinforced, and it also considered more capabilities to work with semi-structured data emerging from the internet usability.

**2009:** University of California, Berkeley researchers developed the Spark project. The DAMA-DMBoK (DAMA International's Guide to the Data Management Body of Knowledge) framework was designed to enable structured data management.

**2010:** James Dixon, Pentaho's CTO, introduced the data lake concept in his blog.[5]

**2011:** Nathan Marz introduced the Lambda architecture through a post in his blog.[6] Google launched BigQuery, a new data warehouse that enables analytics for a large volume of data that performs in the cloud.[7]

**2012:** Snowflake was developed, and the company was founded.[8]

---

[5] James Dixon's Blog (2010), "Pentaho, Hadoop, and Data Lakes." Available at `https://jamesdixon.wordpress.com/2010/10/14/pentaho-hadoop-and-data-lakes/`

[6] Nathan Marz (2011), *How to Beat the CAP theorem: Thoughts from the Red Planet.* Available at `http://nathanmarz.com/blog/how-to-beat-the-cap-theorem.html`

[7] Google Cloud (2011), *Big Query Release Notes.* Available at `https://cloud.google.com/bigquery/docs/release-notes`

[8] Snowflake.com (2011), "How It All Started." Available at `https://www.snowflake.com/pt_br/company/overview/about-snowflake/`

**2013**: The creators of the Spark project donated it to the Apache community and founded the Databricks company. The AWS created Redshift, a new data warehouse that enables analytics for a large volume of data that performs in the cloud.[9] Jay Kreps introduces the Kappa architecture through a post in his LinkedIn profile.[10]

**2014**: Databricks created Apache Spark1 under the governance of the Apache Software Foundation.

**2014-2015**: The DataOps concept became famous among the community due to Chris Bergh's evangelization.

**2015**: Dan Linstedt writes the *Building a Scalable Data Warehouse with Data Vault 2.0.* UC Berkeley researchers developed the Spark project. The DAMA-DMBoK (DAMA International's Guide to the Data Management Body of Knowledge) framework was updated to englobe new needs related to big data.

**2017**: The DataOps Manifesto[11] is created.

**2018**: Databricks created Delta Lake.[12]

---

[9] AWS (2013), "Document history: Amazon Redshift". Available at `https://docs.aws.amazon.com/redshift/latest/dg/doc-history.html`

[10] Jay Kreps (2013), "The Log: What every software engineer should know about real-time data's unifying abstraction," LinkedIn Engineering. Available at `https://engineering.linkedin.com/distributed-systems/log-what-every-software-engineer-should-know-about-real-time-datas-unifying`

[11] `https://dataopsmanifesto.org/`

[12] Michael Armbrust, Ali Ghodsi, Reynold Xin, Matei Zaharia, and Arsalan Tavakoli-Shiraji (2020), "Delta Lake 2.0: Open Source Release," Databricks Blog. Available at `https://www.databricks.com/blog/2022/06/30/open-sourcing-all-of-delta-lake.html`

**2019**: Zhamak introduced the data mesh concept in a blog post.[13] Databricks released Delta Lake as an open source project in partnership with the Linux Foundation.

**2020**: Databricks officially describes the data lakehouse concept on their blog.[14]

The modern data architecture movement emerged to enable any organization to leverage innovations and technologies in a customized manner.

**2024**: Jessika Milhomem introduces the Golden Data Platform architecture design she created to manage data as a product. She also introduces the Data Product Management framework and how to use the Canvas she created to manage any type of product, as you'll see in Part II of this book.😄 🖤

# The Problems

New or existing problems and challenges kept appearing during the data analytics journey.

To better understand how the big data architecture was technically designed, let's start with the problems faced since the early 2000s. (Again, the most relevant topic of this book, how to manage data as a product, is covered in Chapter 5.)

---

[13] Zhamak Dehghani (2019), "How to Move Beyond a Monolithic Data Lake to a Distributed Data Mesh," MartinFowler.com. Available at https://martinfowler.com/articles/data-monolith-to-mesh.html.

[14] Ben Lorica, Michael Armbrust, Reynold Xin, Matei Zaharia, and Ali Ghodsi(2020), "What Is a Lakehouse?" Databricks Blog. Available at https://www.databricks.com/blog/2020/01/30/what-is-a-data-lakehouse.html

- **Even greater necessity to use and manage a large volume of data**: with popular and established social media, new types of businesses started requesting streaming functionalities. It became fundamental to manage a large volume of data in different variabilities for different approaches to analytics.

- **The ability to use data in a timely manner**: Equally important in managing a large volume of data, it's necessary to use analytics data for faster decision-making, such as having updated dashboards with online data to monitor a business operation and investigate any uncommon behavior. Or use the data for online operations, such as a financial transaction risk evaluation. Or even more, to differentiate the business with AI features, such as algorithms to recommend products for a customer during a purchase or discovery moment in an online store, as Amazon. com started doing with book recommendations.

- **Analytics data in operational services with appropriate latency for the business needs**: The latency of the data can create a huge difference in the business, for good or bad, in the same proportion. If the customer uses any online solution that is critical to have a fast answer, such as the risk evaluation and approval of a transaction, it's critical to have a huge latency of the data.

- **Strong concurrences in the market due to globalization**: With large volume and strong participation of transnational companies, using data to get and create the most intelligent and/or innovative solutions is key for businesses.

- **The necessity of all types of data for discovery and analytics**: The same challenges of the past exist but with a bigger proportion. The new, disruptive, innovative approaches and solutions for problems are vital to the companies. Again, those who can better evaluate its data have more opportunities to become more efficient and influential.

- **Full data for analytics and advanced analytics**: The same needs of the past exist but are proportionally more escalated. Having only one perspective of the data is not enough. For a better understanding, the data must be related to the present (most recent information as possible) but also contain historical information.

- **The ability to integrate operational solutions with analytics solutions**: The same need exists for not impacting the operational environment. However, it's fundamental that businesses can use analytics solutions and the operational environment as part of it.

- **Criteria to manage the data life cycle**: Volume and computing capacities are expensive (even with cloud facilities). Thus, creating a good strategy to prioritize and optimize historical data storage is fundamental.

# Big Data

Big data has emerged to evolve the capacity of traditional analytical architecture scalability.

Its main purpose is to store and process large volumes of data, but also to enable analytics and advanced analytics in scale.

The evolution of AI is completely dependent on the data architecture capabilities. For this reason, it's fair to say that although we're in the AI era, we're also still in the big data era. AI doesn't exist without it.

***Figure 3-1.*** *Big data definitions*

Big data started with three V's as its main definitions. It has evolved to five and keeps evolving. This book covers eight of them (see Figure 3-1).

- **Volume**: The capacity to store, compute, and analyze big volumes of data and keep the scalability of it.

- **Variety**: The capacity to compute and analyze different types of data: structured, semi-structured, and not structured (e.g., texts, social relationships, images, videos, etc.)

- **Velocity**: The capacity to ingest, compute, and analyze massive data faster, considering the necessity of the timeliness of the business requests.

- **Value**: The capacity to extract and generate value, regardless of the type of data, at any time (e.g., historical data, recent data, event-driven, raw or treated, statistical, etc.)

- **Veracity**: The capacity to ensure the canonical data is authentic. In short, ensure useful information, not disinformation.

- **Visualization**: The capacity to recurrently monitor, quickly visualize, and give minimum enough information for further analytics or direct final decision.

- **Viscosity**: The capacity to push to action due to the criticality that the a-ha information gave.

- **Validity**: Data is managed with strong governance policies, monitoring, and strategies, ensuring all aspects of data management.

Big data is an ecosystem that enables technical data management focused on analytics, advanced analytics, and data productization for operational services. It is the evolution of the data warehouse.

In terms of architecture, it can be composed of a data lake or data lakehouse, which also coexists with SQL and No-SQL repositories to compose the big data architecture.

## The Start of the Big Data Era

Big data is critical in the AI era. After all, a lot of data is fundamental for AI and needs to be set in a data infrastructure that supports it.

The start of big data happened with revolutionary work by Google. Google was suffering from huge challenges regarding the volume of data and worked on internal projects to solve it. Then, they published papers on the fundamental concepts of Google File System, Map Reduce, and BigTable. These papers shared with the community became the blueprint for sequentially emerging innovations.

Yahoo, also suffering from similar issues, supposedly inspired by the papers, created Hadoop, which was shared with the Apache open source community and enabled the creation of some solutions to address different needs.

Hadoop is an open source framework that distributes, stores, and processes huge volumes of data across a clusterized infrastructure.

It contains different tools that consists of its ecosystem to support the analytics needs, such as its own storage system, computing engine, ingestion, and orchestrator.

Many other solutions emerged and are still emerging in the open source environment. It created additional challenges, as explored in upcoming chapters.

# DW 2.0

In the big data era, three main types of data exist, but exponentially increasing data has become even more relevant and important to consider.

- **Structured**: Data is organized in a tabular schema that contains columns and rows. It's the traditional and easier way to apply analytics (e.g., data from relational databases and spreadsheets).

- **Semi-structured**: Although not having a tabular schema, it is easier to read than unstructured data. It contains metadata, which is fundamental to enabling analytics (e.g., logs, emails, JSON, XML, etc.).

- **Unstructured**: It doesn't have a schema defined. Advanced and robust technical techniques are required to analyze the data (e.g., images, audio, and videos).

In 2008, Inmon introduced DW 2.0 through his book with the same name. In this book, the data life cycle is reinforced, covering the work with semi-structured data, one of the biggest needs derived from internet usability.

After that, some vendors created innovative data warehouse solutions to handle some of these issues, such as Google Big Query, AWS Redshift, and Snowflake.

However, it covered part of it, but not fully. There was still a huge volume of data, and it was necessary to use unstructured data for analytics besides semi-structured data.

During the big data emergence, the main interest was to leverage as much as possible the innovation created so far. Thus, one concept emerged, became popular, and started to be widely used around the world: the data lake.

# Data Lake

This taxonomy emerged in 2010 when James Dixon, Pentaho's CTO, used the data lake nomenclature[15]. After that, the market started to use it for the architecture composed of the products they created for big data.

---

[15] James Dixon's Blog (2010), "Pentaho, Hadoop and Data Lakes." Available at https://jamesdixon.wordpress.com/2010/10/14/pentaho-hadoop-and-data-lakes/

A data lake is a big data repository that contains structured, semi-structured, and unstructured data.

This architecture focuses on the schema on-read approach instead of the schema on-write. In other words, the approach analyzes data by applying business definitions and requirements during the self-service analytics realized through queries on the raw data. Therefore, analyses are fragmented in this approach.

It brought some relevant benefits, including the following.

- **Access democratization and self-service**: Any data user can create their own analytics. By the way, here, you can observe the consolidation of the movement between 2000 and 2010 in the marketing of analyzing fragmented data with the self-service approach.

- **More opportunities for discoverability**: With all data available for usage, data users have even more opportunities.

Machine learning uses all kinds of data, and together with that, more opportunities to evolve AI technologies.

Many other solutions emerged and still emerge in the open source environment, which created additional challenges, which are explored in the upcoming chapters.

## The Components

The creation of Hadoop was the big start to the big data usability around the business world.

It's important to remember that Hadoop is a comprehensive framework composed of a vast ecosystem. During the early steps of the big data journey, many new ecosystem pieces were emerging.

The book doesn't cover all of them, but it's relevant to discuss the most important ones.

- **HDFS** (Hadoop Distributed File System) is the solution for storage management and is distributed in the cluster. Consequently, it has the capacity of storing a huge volume of data.

- **Map Reduce** is the framework to process data in a distributed manner along the cluster, which enables the processing of a huge volume of data.

- **Pig** is a programming language developed by Yahoo to analyze and process large datasets within the Hadoop infrastructure. This was one of the interfaces to work with the data lake for data discovery and analytics.

- **Hive** is a repository used as the interface for the users to interact with the data. This was and still is one of the interfaces (the most common one) to work with the data lake for data discovery and analytics.

- **Sqoop** is a solution that enables data ingestion from relational databases to Hadoop and data loading from Hadoop to relational databases.

- **Flume** is a solution to collect, ingest, and load event data in real-time in a large volume in a distributed environment, such as logs.

- **Mahout** is a lib that contains ML algorithms that can be executed in a distributed environment.

- **Zookeeper** is a solution created to orchestrate configuration management and coordinate Hadoop clusters.

- **Oozie** is a workflow scheduler used to manage the data pipelines created within Hadoop.

Two commercial companies started working with the Hadoop framework, especially in the on-premise environment: Cloudera and Hortonworks, which were recently unified.

The cloud approach has started to become more influential in the market, mainly due to the facilities offered in terms of cost reduction while maintaining the flexibility of growth and escalation of infrastructure.

Thus, players such as AWS, Google, Microsoft, and others started to create their own solutions and offer them through cloud services. It was especially relevant for startups, becoming the most relevant and used infrastructure approach nowadays in the market.

## Similarities with the Traditional Data Architecture

The interesting thing about analyzing the characteristics of the data lake repository in comparison with traditional architecture (besides the obvious similarity with the data warehouse as a repository to apply analytics—even without the same definitions and characteristics) is that it also identifies similarities with the staging area layer.

Although the staging doesn't contain the variety characteristic of big data, the rest is pretty similar. It contains the data from the original sources in a centralized repository, with raw data, and is used for analytics. However, it has adapted and evolved to the new necessities of the market: big data characteristics and enabling democratized access for any kind of user. It enables them to use and explore the repositories through analytics.

This similarity is one of the reasons I see it is so important to understand and not underestimate the history of data architectures. We faced similar needs in the past, so we can adapt to the current reality and address or mitigate new problems.

While big data gained more users, new problems and challenges were identified.

One is related to the difficulty of understanding which data can be used as trustful data, as you'll see further in the next sections. Another fundamentally important one is having fresher and faster data to be used in operational activities for the business and intelligent functionalities and activities. The full data is refreshed with historical, but late hours of freshness are not the only need for the business. Besides, although not completely new, we had the ODS in the past, but now there is massive data to be analyzed and used for analytics.

Data is already part of the business, not a differential. It's vital for existence.

# Lambda Data Architecture

Nathan Marz introduced the Lambda architecture in 2011, and in 2015, Marz and James Warren developed it further[16].

The Lambda architecture is a data deployment model that combines batch[17] and streaming[18] processing (real-time[19] and near real-time[20]) to handle large amounts of data. In other words, this architecture was designed to handle "big data."

---

[16] Nathan Marz and James Warren (2015), *Big Data: Principles and Best Practices of scalable Real-Time*, Manning Publications.

[17] Batch data is a computer method programmed to collect and process data points together periodically, within a specific period of time, considering historical and new updated data as reference. It's more out-of-date from the real-time. Commonly it takes hours to be executed, and the time window of data consists of a specific start date (which can be full or incremental, by using deltas) and ends in the D-1 (day before the current one). However, it can also execute in any window of hours or minutes, lower than 24h. When the incremental data processing is used and the delta end date/hour is lower or equal to 5h, it's considered a micro-batch.

[18] Streaming is a flow of data that consumes diverse sources and aims to have data consumed in real-time and/or near real-time.

[19] Real-time data, in the opposite of the batch data, is the data made available to use as soon as it is generated. There's no common definition for the limit of latency of it, but I'd say it takes less or equal to 0.5 minutes for usage.

[20] Near real-time is similar to the real-time data, however, it requests more transformations before making it available for use. There's no common definition for the limit of latency of it, but I'd say it takes less or equal to 5 minutes for usage.

# Lambda Architecture's Data Layers

The Lambda architecture has three main layers.

- **Batch layer**: This is where the common and traditional process for the data used for analytics occurs. It processes data consumed from diverse sources (immutable raw data), considers a period of time of data (historical and new updated data), applies transformations, and stores the final data as a batch view in the serving layer for consumption.

    Since the data is outdated compared to real-time data, it's also denominated the cold path. I'd say it is warm, as most business processes commonly use it, and cold data is related to more unused data, usually stored in specific repositories for this type of use.

- **Speed layer**: Opposite the batch layer, the goal of this layer is to process and store data for analytics in the serving layer with near/real-time information as speed layer views. For this reason, it's designed to support lower latency. It's a hot path because it works with near/real-time data.

    The counterpart of the short latency is the accuracy of the data, as it contains just a piece of the data perspective: the most recent one, not the full one, as the batch layer.

- **Serving layer**: This layer contains data for consumption. It can be a common serving layer, which merges both data (batch + speed), or it can have dedicated serving layers where the data is stored separately for consumption.

# Architecture's Characteristics

- **Architecture**: Clear layers facilitate the usability of the data, with clear scope, resolutions, and challenges. Consequently, it creates definitions of the anti-corruption layer for the data.

- **Use case**: It's the ideal architecture for scenarios of analytics, where data must be retained and analyzed together with historical data, such as traditional analytical needs (e.g., credit risk analytics, customer segmentation, fraudsters, and suspect detection).

  However, it also attends scenarios where it's necessary to have just the online data without the immediate need for historical data (e.g., IoT sensors, fraudulent transactions detection, market trends monitoring). Both abilities are differentiated.

- **Performance**: Good performance in executing processes and usability of the data, as they have specific goals and paths for usage.

- **Adoption**: Adopting is easier because half of the architecture is generally already in place for the institutions, making the adoption easier with the batch perspective. Then, the new designs, implementations, and adaptations remain in the speed layer.

- **Coding and maintenance**: Although it's possible to reuse many logics by eventually leveraging them, it contains two systems requiring different paths for coding, one for batch and another for the speed layer. Consequently, maintenance is required more.

- **Cost management**: The speed layer focuses on a specific period of events for processing and the availability of data for usage, and the batch layer can be an expensive approach. This is because it'll request the reprocessing of full historical data if the Delta Lake features are not used.

# Kappa Data Architecture

Jay Kreps introduced the Kappa architecture in 2013[21] to simplify the Lambda architecture using the same structure as the Lambda architecture; however, there is just one data flow path: the stream processing approach.

## Kappa Architecture's Data Layers

The Kappa architecture has two main layers focused on the hot path.

- **Streaming layer**: It uses just one engine: stream processing. The goal of this layer is to support lower latency for immediate insight. The stream events are ingested as soon as they are generated and consolidated into a unified log. Then, the data is processed by ordering the events and appending new events. Finally, the data is stored in the serving layer.

- **Serving layer**: The layer that contains the data for consumption. All the events are processed since the data is ingested and persists as a real-time view.

---

[21] Jay Kreps (2013), "The Log: What every software engineer should know about real-time data's unifying abstraction," LinkedIn Engineering. Available at https://engineering.linkedin.com/distributed-systems/log-what-every-software-engineer-should-know-about-real-time-datas-unifying

# Architecture's Characteristics

- **Architecture**: Simplified data layers that use one unique approach for processing data.

- **Use case**: Ideal for typically online use cases that avoid historical data (e.g., IoT sensors, fraudulent transactions detection, market trends monitoring).

  If necessary to use batch data, it's possible to consume the data from the cold path, however, and replay the stream. However, finding availability may take longer than common streaming data and become a micro-batch.

- **Performance**: Good performance in executing processes and usability of the data if used just for streaming data. However, it can have a performance impact due to the usage of warm and even more for cold data, if necessary.

- **Adoption**: Although it requires fewer resources than the Lambda architecture and is more simplified, it is complex to adopt by existing companies, as it'll be necessary to implement new analytical engines and technologies.

- **Coding and maintenance**: Although the implementation is easier, due to usually not having duplicated code as its purpose, it may deal with duplicated events and sequences that require additional effort, especially to keep it executing in the expected requirements for online, which is the proposal of this architecture.

137

- **Cost management**: It's an expensive architecture just when it's necessary to consider the usage of warm or cold data in the processing because it can request recurrently reprocessing of data, if not using the Delta Lake features.

# Medallion Data Architecture

Although it's very relevant, the medallion data architecture is commonly underestimated by professionals in the industry.

Please keep it in mind because, besides being relevant, it is a key part of the proposal of this book.

That being said, before talking about its definitions, let's recap that in the initial era of big data, the main approach was to data wrangle the data just at its usage.

This means everybody who worked and needed data used the same approach. Companies born in the big data era are already growing in this new environment. At the same time, the existing companies were working with two environments, a legacy and a new one, in the middle of a huge data architecture migration.

As explained in the data lake section, this approach created advantages. It solved some issues, such as the self-service analytics that accelerated the operational processes that required discoveries and explorations directed by data users.

However, in terms of data quality, it was a regression, as the companies lost control of data veracity from a global perspective. Many operational processes evolved with this democratic process, but many others that were already solved with the traditional analytics approach regressed.

Important to reinforce a disclaimer here. It was a common symptom in the market when a new solution emerged and became a buzzword. New means revolutionary and better. The traditional solutions are obsolete and don't work anymore. Clearly, it is not an absolute truth.

In my vision, the issues related to the data veracity were a result of a complete disinterest of new professionals in the data world to understand and learn about the data and the full context of the business and history, and from the more senior ones that underestimate the full journey of data, the issues, challenges, and historical information (when they had enough knowledge).

Again, everything is important and relevant, and it is the responsibility of a manager of data products to keep studying, learning, and creating technologies, but always focused on the business needs and existing and new potential issues. The focus must be on problems and the domain context, and upon them, design products and solutions, whether new ones or picking the best components to create a customized one.

Returning to the early stages of the big data era, the necessity of a solution for this scenario was clear. Thus, the medallion architecture was introduced by Databricks in 2018, aiming to clearly define the layers of canonical data (a.k.a., source of truth) for analytics usability. This architecture has three main types of data layers.

## Bronze Data

This layer comprises raw data, meaning it comes from operational sources, regardless of the type of data: structured, semi-structured, or non-structured. As a characteristic, it has little or no data treatment.

When applied, it has minimum data wrangling and is focused on pretreating the data for analytics usability (e.g., transforming a JSON into a tabular format).

## Silver Data

This layer is composed of treated data, but it doesn't have all the requirements to be considered federated data to be consumed by anyone and any process.

I see this type of data exists for two purposes.

- Created upon an exploratory scenario (e.g., A/B tests, where business rules are applied to test some product, product feature, and/or customer behavior)

- As part of a process to refine and transform golden data (e.g., datasets with intermediate business rules, which alone don't answer much but compose a complete accounting reconciliation)

## Golden Data

This layer comprises a source of truth data that must be used for any perspective of the business context: inside the business domain, by other business domains that need to use their own context, or to share the data with external entities.

It is composed of bronze or silver data and has a federated vision.

It contains all the data management and governance requirements applied.

A good example is the accounting reconciliation data used internally by product teams and horizontal teams, such as FPnA, controllership, and risk, and also shared with the regulatory agencies due to laws and regulations.

I'd say this layer is not related to the granularity of the data because it should fit any type of granularity. As Databricks describes,[22] it fits Inmon's or Kimball's data modeling approaches for data marts.

To me, the key definition for golden data ensures the availability of data with strong veracity, value, and validity, some of the V's defining big data.

---

[22] Databricks.com (2013), "What is a Medallion Architecture?" Available at https://www.databricks.com/glossary/medallion-architecture

As you can see, there are many similarities with the proposal of the data warehouse.

# Data Lakehouse

The data lakehouse emerged due to the need for a more organized architecture to enable analytics.

Although already used in the market (such as the Redshift Spectrum, created by AWS in 2019 and called a lakehouse), this taxonomy was coined by Databricks,[23] which raised it in 2020 on their blog and further evolved it with Delta Lake.

Databricks describes a data lakehouse as a combination of data lake and data warehouse features, especially considering ACID features.[24]

I'd adjust its definition by removing the obligation of ACID functionalities, mainly because although these functionalities are super relevant and beneficial in many aspects of the data architecture, it's unnecessary to have ACID functionalities to ensure the analytical data architecture.

A data lakehouse is the data management architecture that combines the data warehouse and data lake's best features to enable any analytics and advanced analytics data.

# The Combination

A data lakehouse is an architecture that combines a data lake with data warehouse ecosystems purposes.

---

[23] Ben Lorica, Michael Armbrust, Reynold Xin, Matei Zaharia, and Ali Ghodsi (2020), "What Is a Lakehouse?" Databricks Blog. Available at https://www.databricks.com/blog/2020/01/30/what-is-a-data-lakehouse.html

[24] Databricks.com (2020), "What is a Data Lakehouse?" Databricks. Available at https://www.databricks.com/glossary/data-lakehouse#:~:text=A%20data%20lakehouse%20is%20a,(ML)%20on%20all%20data

Consequently, it contains, supports, and implements the following.

- All types of data for any analytics purposes (bronze, silver, and golden data)

- Any type of data timeliness (batch and streaming data)

- Any type of analytics and advanced analytics with any type of data (unstructured, semi-structured, and structured)

- Applications with analytics purposes

Data management enables complete data management of the architecture, ensuring the governance for the whole stack.

## Consolidating and Moving to the Next Steps

This approach observes the marketing movement that started in 2010 and lasted a decade for a more unified data analysis.

In terms of detailed design, although it can also be implemented with the Kappa architecture, it fits more with the Lambda architecture, as, by definition, it contains the batch layer, a super relevant piece for the data warehouse ecosystem needs and characteristics.

Nonetheless, we cannot forget that there are still challenges related to batch layer processing, especially related to the flexibility for cost-efficiency, which creates limitations for scale. For this reason, Delta Lake was created.

## Delta Lake

One of the biggest issues the market has been facing in the data lake is related to the limitations for data manipulation within big data architectures.

The limitation requests the full reprocessing and persistence of historical data in a batch. This means that, for batch, every day, all the data of institutions have to be reprocessed. It created many issues for the companies, such as the long time to execute the pipelines, which consequently generates delays in the availability of warm data for analytics, huge costs, and impacts on the business for business processes that use that data.

In this scenario, the Delta Lake emerges. It allows ACID transactions (atomicity,[25] consistency,[26] isolation,[27] durability[28]) and other data reliability functionalities from the data warehouse ecosystem to be performed via Spark during data processing.

# Brief Introduction to Apache Spark

Before discussing Delta Lake's characteristics, let's quickly cover Spark.

Apache Spark is an open source analytics engine that processes large amounts of data in near/real-time or in batches.

It started as a project studied and performed by researchers at the University of California, Berkeley, in 2009.

In 2013, the creators of the Spark project donated it to the Apache community and founded the Databricks company. Since then, Spark has evolved and become the most used technology worldwide to process big data.

---

[25] Atomicity: The entire transactional takes place at once or doesn't happen at all.

[26] Consistency: The database must be consistent before and after the transaction.

[27] Isolation: Multiple transactions execute in an independent manner and do not impact each other.

[28] Durability: A successful transaction is committed and persisted in the database, even if there are system failures.

# Delta Lake's Characteristics

The Delta Lake project began in 2018 using Apple as a use case for some specific scenarios, and eventually, it expanded to their customers.

In 2019, Databricks released Delta Lake as an open source project in partnership with the Linux Foundation.[29]

The following are the most relevant Delta Lake characteristics.

- It enables ACID transactions.

- It enables data manipulation language (DML) commands, such as updates and deletes.

- It enables schema enforcement.

- It can be performed by many frameworks besides Spark, such as Hive, Flink, and Kafka.

- It integrates with many managed services, such as AWS Athena and AWS EMR.[30]

The market is still learning this functionality. Since it became open source, other solutions, such as the Apache Iceberg, emerged in the market and possibly kept emerging and evolving.

---

[29] Michael Armbrust, Ali Ghodsi, Reynold Xin, Matei Zaharia, and Arsalan Tavakoli-Shiraji (2020), "Delta Lake 2.0: Open Source Release." Databricks Blog. Available at `https://www.databricks.com/blog/2022/06/30/open-sourcing-all-of-delta-lake.html`.

[30] `https://delta.io/integrations/`

# DataOps

DataOps is a shortened version of *data operations*. It emerged in the mid-2010s. Lenny Liebmann first mentioned it in a 2014 post titled "3 reasons why DataOps is essential for big data success" on IBM's blog, and L. Ganesh Padmanabhan discussed it in a blog post in 2015.[31]

It became popularized by Chris Bergh, the CEO of DataKitchen, who evangelized the community during the same period and is mainly credited for creating The DataOps Manifesto[32] in 2017.

A more common definition of DataOps is that it's a methodology that leverages DevOps, agile, and lean manufacturing to manage data through automation, monitoring, and collaboration among professionals while focusing on efficiency improvement.

I also like Michelle Goetze's[33] definition, in which she points out that DataOps is the ability to enable solutions, develop data products, and activate data for business value across all technology tiers, from infrastructure to experience.

I believe DataOps is fundamental for implementing data products. And the Golden Data Platform leverages DataOps functionalities, as you'll see in Chapter 5.

DataOps is grounded in five values.

- Individuals and interactions over processes and tools

- Working analytics over comprehensive documentation

- Customer collaboration over contract negotiation

---

[31] L. Ganesh Padmanabha (2015), "DevOps to Data Ops," DataScienceCentral.com. Available at https://www.datasciencecentral.com/from-devops-to-dataops/

[32] https://dataopsmanifesto.org/

[33] Michelle Goetze (2020), *DataOps for the Intelligent Edge of Business*, Forrester. Available at https://www.forrester.com/report/DataOps-For-The-Intelligent-Edge-Of-Business/RES162717

- Experimentation, iteration, and feedback over extensive upfront design
- Cross-functional ownership of operations over siloed responsibilities

The DataOps manifesto principles are as follows.

- Continually satisfy your customer through early and continuous deliverables.
- Value working analytics.
- Embrace change to generate competitive advantage.
- Know it's a team sport with diversity for better innovation and productivity.
- Daily interactions with customers and all involved and interested ones.
- Be self-organized.
- Reduce heroism and create sustainable scalability.
- Reflect on continuous improvements.
- Analytics is code.
- Orchestrate all solutions and types of them.
- Make it reproducible.
- Disposable environments through easy and cheaper creation.
- Simplicity.
- Analytics is manufacturing.
- Quality is paramount.
- Monitor quality and performance.

- Reuse manufacturing.

- Improve cycle times.

DataOps involves business requirements (the Data perspective) and business value (Ops perspective) (see Table 3-1).

*Table 3-1. DataOps Processes*

| DATA | OPS |
|---|---|
| • **Plan** the solution by mapping the business requirements. | • **Release** the data solutions to ensure the delivery of business value. |
| • **Develop** the data solution considering the business requirements. | • **Deploy** the data solution to achieve the business value. |
| • **Build** the data solution. | • **Operate** the data solution to ensure business value. |
| • **Manage** the data solution. | |
| • **Test** the data solution considering all business requirements. | • **Monitor** the data solution to ensure continuous business value. |

DataOps leverages the lean manufacturing concept by focusing on increasing the efficiency of ingesting/consuming, processing/computing, storing, and analyzing the data while maximizing the value of the deliverable.

DataOps is focused on ensuring the following.

- **Automation** is a key principle.

- Take advantage of **tools and standards**, including and especially open source technologies.

- Work through **data contracts** for clear and trusted integration and agreements between producers and consumers of data.

# Data Contracts

This concept started to be discussed with the DataOps concepts, and since 2022, it has been popularized by people in the data community evangelizing the concept.

Data contracts are terms and conditions defined between producers and consumers of data, with automation to ensure the data is transmitted accurately and consistently.

Data contracts had gained more attention by 2022 when Andrew Jones's blog[34] became famous among the data community. However, the practice has evolved since its emergence. Some technologies are open source, such as the dbt framework (data built tool), which works with contracts for schema, data quality controls, and configurations.

# Data Consumption and Data Production

Before discussing data contracts further, it must be made clear that they can be defined for data consumption and data production.

Nice, but what exactly does that mean?

Excellent question! I've heard from senior leaders, product managers, and specialists in different and admired companies that data production is just realized by services, and data ingestion is not data consumption. Because of this kind of mindset we have been facing such painful processes for data quality, also to work and evolve with data. And most importantly: the reason we have so many customers frustrated and septic with the data they consume.

Thus, let's make it clear: there's always data consumption and data production. Always. It can occur at different moments and in different

---

[34] Andre Jones (2022), "Data Contracts at GoCardless — 6 Months On," *Medium*. Available at https://medium.com/gocardless-tech/data-contracts-at-gocardless-6-months-on-bbf24a37206e

aspects but remains as production and consumption. And no, data production is not realized just by services. It is also realized by analytics environments. Both must be treated as their aspects and responsibilities.

## Data Consumption

Data consumption is a process related to an analytical purpose. It can be realized from structured, semi-structured, and unstructured data.

It can be realized through a data ingestion process, which is the process of collecting and storing or moving the data to a repository. Thus, it means that the data originates from transactional/operational environments and platforms (e.g., events from services (transactions from users, logs, images, videos).

Or it can be realized through business intelligence (BI) or advanced analytics tools. Thus, the data originates from analytical architectures (e.g., IDE for queries, notebooks, dashboards, and user interfaces).

Or it can be realized through APIs, with data originating from transactional/operational platforms (e.g., transactions from customers, logs) or analytical architectures (e.g., ML models execution, batch or real-time data for reports).

## Data Production

Data production is the process of creating or generating data, usually for analytics purposes, while operational/transactional platforms also generate it. It can produce structured, semi-structured, or unstructured data.

It can be performed through transactional/operational environments and platforms with main interaction from users or automation—using AI or not (e.g., transactions from users regarding sales or purchases, transaction operations from customers, creation of logs, generation of images, generation of videos).

Or it can be realized through business intelligence or advanced analytics processes, via automation realized in analytical environments (e.g., batch or streaming processes (a.k.a., ETL processes), ML models outputs from interaction with new data received (from batch or streaming data).

Next, let's move on to data contract types.

# Types of Contracts

There are a few types of data contracts.

### Schema Definitions

The responsibility of a schema contract is to explicitly define the structure of the generated data. If the schema doesn't match the specified structure, attributes, data types, constraints, type of materialization (such as ephemeral or not), and type of request (such as build, rebuild, drop), it fails.

For maintenance, a resolution is possible during the continuous integration, when all the errors related to the failures can be caught through logs.

It is usually implemented via semi-structured files with the metadata definitions, such as YAML. However, it's not limited to it. It can be done via other approaches that ensure automation with standards, best practices, and security, which are important goals of DataOps.

getdbt.com provides an example of a contract defined via a YAML file.[35]

## Data Quality Controls

Data quality controls are responsible for validating the data by checking the general data checks necessary for data quality (or detailed business checks).

---

[35] getdbt.com (2025), "Models Contracts," dbt Developer Hub. Available at https://docs.getdbt.com/docs/mesh/govern/model-contracts

- **General data quality**: General checks are checks
  related to the consistency of the data, such as duplicity,
  nunability, and others. It's usually related to the schema
  check also. It can be implemented via a file such as
  YAML or via abstractions (see Figure 3-2).[36]

```
dbt_purchases_project_processed_schema.yml
1    version: 2
2
3    models:
4      - name: analytics_offers
5        description:  "The offer analytics table, which contains the consolidated result for offers"
6        tests:
7        - unique:
8            column_name: "concat(category, company, brand)"
9
10       - not_null:
11           column_name: "concat(category, company, brand)"
12
13
14     - name: analytics_transactions
15       description: "The transaction analytics table, which contains the analytical transaction data"
16       tests:
17       - not_null:
18           column_name: "concat(id, category, company, brand)"
```

***Figure 3-2.*** *Image captured from the dbt-pipeline use case (part1):*
*Data Transformation blog post from Jessika Milhomem to represent a*
*data contract generated through dbt for general check of schema*

- **Business checks**: This type of contract is responsible
  for evaluating the behaviors of the data that accurately
  translating the expectations of the business rules.

  - Only customers not on the block list should be
    considered for a specific campaign.

---

[36] Jessika Milhomem (2021), "dbt-pipeline use case (part 1): Data Transformation,"
*Medium.* Available at https://jessikamilhomem.medium.com/case-de-
pipeline-com-dbt-parte-1-transforma%C3%A7%C3%A3o-de-dados-a7dc9dfbb165
(in Portuguese) or https://jessikamilhomem.medium.com/dbt-pipeline-use-
case-part-1-data-transformation-396fae8275de (in English)

- A customer cannot be tagged as late if the due date of bills is earlier than the bill payment date.

- Consider a deviation from a specific threshold.

It is usually implemented via semi-structured files, such as YAML. However, it's not limited to it. It can be done via other approaches that ensure automation with standards, best practices, and security, which are important goals of DataOps.

Soda.com provides an example of a contract defined via a YAML file.[37]

## Security and Privacy Restrictions

The responsibility of the security and privacy restrictions contract is to specify the security and privacy restrictions and ensure the data as a product complies with security policy definitions. The organization defines the security policies but must also include and comply with the appropriate data privacy and protection regulations, such as GDPR, HIPAA, PCI DSS, and LGPD.

It can be implemented via a file or via abstractions.

getdbt.com provides an example of a contract for public or private data.[38]

Soda.com provides an example of a contract defined for PII data.[39]

[37] Soda.com (2025), "Write SodaCL Checks." Available at https://docs.soda.io/soda-cl-overview

[38] getdbt.com (2025), "Model access," dbt Developer Hub. Available at https://docs.getdbt.com/docs/mesh/govern/model-access

[39] Soda.com (2025), "Write SodaCL Checks." Available at https://docs.soda.io/soda-cl-overview

## Versioning

The responsibility of a versioning contract is to ensure the versioning of the data definitions. It's usually related to the schemas but can be defined apart and for different needs.

It is usually implemented via semi-structured files. However, it's not limited to it. It can be done via other approaches that ensure automation with standards, best practices, and security, which are important goals of DataOps.

getdbt.com provides an example of a contract defined for the visibility of public or private data.[40]

# Service-Level Agreements

The responsibility of a service-level agreement (SLA) contract is to define the commitments about the availability and freshness of data. Data consumers need to ensure confidence and efficiency.

It includes commitments related to data completeness, data failure recovery, and data freshness.

# Shift-Left Testing

Shift-left, also known as shift-left testing, is a software testing approach created by Larry Smith. He wrote a post called "Shift-Left Testing" in 2001[41] that proposes detecting and fixing issues as early as possible in the development process. Thus, the testing is anticipated.

---

[40] getdbt.com (2025), "Model versions. dbt Developer Hub." Available at https://docs.getdbt.com/docs/mesh/govern/model-versions

[41] Larry Smith (2001), "Shift-Left Testing," Dr Dobb's. Available at https://www.drdobbs.com/shift-left-testing/184404768

Although it has been massively discussed in the data community recently, this approach has been leveraged as part of DataOps, mainly applied for data quality monitoring.

Its main goal is to evaluate and fix as many as possible upstream issues, closer to the source, instead of keeping it just for the downstream.

The practical approach is to use data contracts to apply it in the data pipeline. It can also be applied to the schema or data quality validation.

It can be implemented via a file such as YAML or via abstractions.

For example, the dbt framework requires the configuration of source files, where the source must be configured. Similar schema configurations and data validations can be applied to this contract (see Figure 3-3).[42]

```
dbt_purchases_project_purchases.yml

1    version: 2
2
3
4    sources:
5      - name: warehouse
6        tables:
7          - name: transactions
8          - name: offers
```

*Figure 3-3.* *Image captured from the dbt-pipeline use case (part1): Data Transformation blog post from Jessika Milhomem to exemplify a data contract used for sources checks through dbt*

---

[42] Jessika Milhomem (2021), "dbt-pipeline use case (part 1): Data Transformation," *Medium.* Available at https://jessikamilhomem.medium.com/case-de-pipeline-com-dbt-parte-1-transforma%C3%A7%C3%A3o-de-dados-a7dc9dfbb165 (in Portuguese) or https://jessikamilhomem.medium.com/dbt-pipeline-use-case-part-1-data-transformation-396fae8275de (in English)

# Data Warehouses and Data Marts

In the era of big data architecture, data warehouses and data marts are still in use; however, now, there are additional possibilities to develop an architecture that is more aligned with the needs of the business.

Due to these innovations and different possibilities, modern data architecture is a new movement.

Modern data warehouses are also able to work with semi-structured data, and companies working with an enormous amount of data can use these new technologies. Companies with all the big data characteristics, except for the huge amount of data, such as startups or specific segments, can also leverage it.

The reason for this is the creation of data warehouses capable of performing as a data lake in its infrastructure, with modular data infra components for management.

Thus, there are different possibilities for working with data warehouses and data marts in this equation, considering some specifics. Let's look at the possibilities next.

## Data Warehouses and/or Data Marts Managed in Different Infrastructure Platforms

The approach here is to have the data warehouse and/or data marts outside the main infrastructure platform that enables the creation and maintenance of all capabilities.

- **Data warehouses composed of data marts**: They follow Kimball's design vision, where the data marts consists the data warehouse. The infrastructure it defines is done with a repository of a different infrastructure platform, which can be on-premise, although it is usually in the cloud (e.g., of market products: Google Big Query, Snowflake, AWS Redshift, and others).

- **Data warehouses separate from data marts**: It follows Imnon's design vision, where the data warehouse has a different repository from the data marts. It can be implemented within a similar infrastructure as Hadoop, including data warehouse definitions, with another repository in another infra just for the data marts.

  Or both architectures separated from the data lake infrastructure and managed in different or similar infrastructure repositories.

## Data Lakes, Data Warehouses, and Data Marts in a Unique Infrastructure Platform

The approach here is to have one unique infrastructure platform that enables the creation and maintenance of all capabilities.

- **Large volumes of data** can be collected in cloud or on-premise environments similar to Hadoop, such as AWS EMR and Google Cloud Dataproc. It can also be implemented for similar purposes but with different infrastructure components.

- **Small and medium volumes of data** can be performed on-premise, but it is often performed in cloud environments using vendor solutions, such as Snowflake and Redshift+Spectrum.

In terms of data modeling, mainly two designs have been considered: dimensional modeling, according to Kimball's design architecture, and data vault, according to Dan Linstedt's architecture modeling, which is more aligned with Inmon's design architecture.

The data architecture design and strategy decisions must consider different perspectives of business requirements besides the technical characteristics of the organization.

This book won't go deeper into these details because it's not the goal of the book. But here, we can see another reason for the importance of having a strong data management approach that must be performed by senior leadership to design the complete strategy, create the tactical plan, and be able to put everything in place.

For this reason, it is fundamental to have clarity about so many contexts shared in this book. By the way, let's talk a bit more about ETL, but now in the big data architecture.

# 3.2. Data Ingestion, Data Processing, and Data Persistence

## Approaches

ETL and ELT are two approaches to ingesting, processing, and loading data from traditional analytics architecture. However, the second one, the ELT approach, started to be strongly used, especially after the emergence of big data. That's the reason I decided to include it in this chapter.

We don't define the process for it; we define the architecture to implement it.

- For data lakes, the ELT approach is primarily used.

- The ETL approach is primarily used for data warehouses and data marts (regardless of their architecture design, including data lakehouses).

Let's talk about their general characteristics!

# ETL

ETL is the acronym for extraction-transformation-loading. It is designed to feed data warehouse and data lakehouse architectures.

ETL feeds and supports BI and AI applications because it receives application data and exits enterprise analytical data.

The following describes the main characteristics of ETL (see Table 3-2).

*Table 3-2.* *ETL vs. ELT*

|  | ETL | ELT |
|---|---|---|
| **ACRONYM** | extract-transform-load | extract-load-transform |
| **DATA FLOW** | 1. Data is ingested from the sources. | 1. Data is ingested from sources. |
|  | 2. Data is wrangled: cleaned, business rules applied, and transformed. | 2. Raw and original data is loaded into the repository. |
|  | 3. The final treated data is loaded into the final repository. | 3. Data is transformed as soon as it is consumed. |

(*continued*)

*Table 3-2.* (*continued*)

|  | ETL | ELT |
|---|---|---|
| **STRENGTHS** | • Ideal for batch and backfill data processing and loading<br>• Generates structured data<br>• Ideal for cleansing and transformations<br>• Delivers treated data<br>• Ideal for complex scenarios with rules and transformations<br>• Ideal for bulk data movements<br>• Enables good data quality controls for automated processes<br>• Easier maintenance<br>• Lower ongoing costs<br>• Simpler data privacy compliance with automated taggings and transformations | • Ideal for processes that do not require transformed data for consuming<br>• Ideal for scenarios that request discoveries or simple processes<br>• Ideal for sandbox environments<br>• Fast speed<br>• Move data to the destination more quickly for faster availability<br>• It can work with all data types: structured, semi-structured, and unstructured<br>• Lower costs upfront<br>• Do not require business knowledge, except on the transformation on-read moment<br>• Easy maintenance of individual processes |

(*continued*)

*Table 3-2.* (*continued*)

| | ETL | ELT |
|---|---|---|
| **WEAKNESSES** | • Slow speed<br>• Support just structured and semi-structured data<br>• May request server scaling<br>• High maintenance effort<br>• Necessary business knowledge for implementation and maintenance<br>• Not ideal for sandbox environments<br>• Higher costs upfront<br>• Lower flexibility for data cost reduction | • Does not store data cleaned and transformed<br>• Higher ongoing costs<br>• High maintenance effort when many processes are created with schema-on-read transformations<br>• Never delivers treated data<br>• Not ideal for recurrent processes<br>• More complex for data privacy compliance due to the necessity of automated tagging and transformations |
| **EXAMPLES OF USE CASE** | Data warehouses, data marts, and data lakehouses<br>Streaming data to feed analytical applications that require transformed data (e.g., products/ services recommendation | Stagings<br>Data lakes<br>Streaming data to feed analytical applications that do not require transformed data for consumption (e.g., logs) |

# Speed

- It is slow due to the process of getting the data and transforming it to then load the data treated. For instance, creating data warehouses and/or data marts takes months to produce.

- It's the ideal process to manage large amounts of data, such as backfill data processing and batch processes.

# Data Types

- It can work only with structured and semi-structured data.

- Its main goal is to generate structured data. It pulls data from the legacy system environment and transforms it into corporate data.

- The ideal process to apply cleansing and transformations (e.g., logical conversion of data, merging of records, deduplication of data, etc.).

# Data Complexity

- Always delivers treated data

- Ideal for complex scenarios of analytical data that require complex rules and transformations

- Ideal for bulk data movements

- Request server scaling depending on the volume of the data to be processed and ingested

- Enables good data quality controls for monitoring automated processes

# Maintenance

- It requires a high maintenance effort due to the effort needed to map and implement a process of changing requirements.

- It is easier to maintain due to traceability.

- It is necessary to have business knowledge and, ideally, be business-oriented to implement and maintain the process.

- It is not ideal for sandbox environments like ML modeling.

# Expertise

- Requires business-oriented professionals

# Cost

- The costs are higher upfront due to the characteristics of the environment, a large capacity, and more powerful servers to operate.

- Besides the licensing costs of the ETL software and the expert engineers who can effectively use it.

- It usually has lower ongoing costs.

- It has lower flexibility for large amounts of data cost reduction due to the necessity of hiring powerful servers since the beginning, especially when the data warehouse/data marts are stored in massively parallel processing (MPP) repositories.

## Data Privacy

It may make data privacy compliance simpler due to the automated tagging and transformations.

# ELT

ELT is the acronym for extraction-loading-transformation. It is designed to copy data from the source and load it at the target repository with the as-is data (raw data). Then, it can be transformed as necessary afterward (a.k.a. the schema-on-read process).

This approach can feed traditional architectures (staging areas) because it was created for a traditional data architecture. However, it's more often used for data vault and data warehouse modeling and feeding data lakes with massive data on a large scale.

The following describes the main characteristics of ELT (also see Table 3-2).

## Speed

- It is ideal for processes and applications that do not require transformed data for consumption.

- It is faster due to the lack of necessity for data transformation.

- It moves data to the destination more quickly for faster availability.

## Data Types

- It can work with all data types: structured, semi-structured, and unstructured.

- It does not store data cleaned and transformed.

# Cost

- The costs are lower upfront due to the characteristics of the environment in which it is used.

- It usually has higher ongoing costs.

# Maintenance

- It does not require business knowledge, except for transforming on-read.

- It is ideal for sandbox environments, as the ones for ML modeling.

- It offers easy maintenance of individual processes due to their simplicity.

- It requires a high maintenance effort when many processes are created with schema-on-read transformations.

# Expertise

- Do not require business knowledge, except for transforming on-read.

# Data Complexity

- Never delivers treated data

- Not ideal for recurrent processes

- Ideal for scenarios that request discoveries or simple processes

## Data Privacy

- Data privacy compliance is more complex due to the necessity of automated tagging and transformations.

# 3.3.  Limitations

Although many issues were solved with big data, challenges remain.

The main goal of this section is to understand the issues and limitations faced in the past, learn about them, relate them to our current scenario, and eventually, predict some issues we'll face. Why? Because they are directly related to the development of data products and data as a product.

For this reason, let's discuss some of them by their divisions: storage, computing, and data governance.

Chapter 5 touches on some of the challenges.

# Storage Limitations

Since the creation of the big data architecture, there has been a:

**Large volume of data and management storage issues.**

- **Ability to handle comprehensive data storage**: With the evolution of technologies and the generation of data, the issues in the traditional data architecture of consuming semi-structured and unstructured data have been solved. Technologies like HDFS were fundamentally created for Hadoop, influencing the creation of other solutions to solve this type of issue, such as AWS S3. Data warehouses also have functionalities that enable the storage of semi-structured data, such as AWS Redshift, Google Big Query, and Snowflake.

- **Scalable storage infrastructure**: The big data strategy of using horizontal infrastructure instead of vertical infrastructure (individual servers with high capacity) solved the necessity of powerful machine resources to support the exponential growth of data. In other words, the data persistency is distributed using clusters composed of nodes (computers used as servers). It now works for data warehousing repositories and big data architectures.

- **Persistency of data according to the business use case**: The Lambda and Kappa architectured solved the necessity of using fresher data for analytics and advanced analytics. The Lambda architecture addresses the two main scenarios: batch and streaming data.

  - **Batch data** manages the full historical to support any analytics or advanced analytics that request accurate data. The repositories used for this purpose are addressed by distributed file systems for big data environments, new data warehouse repositories, or via No-SQL databases, depending on the requirements defined for the business use case.

    - **Fault tolerance**: This capability is fundamental for high availability in critical environments. Due to the distributed storage approach, it can maintain proper operation despite failures in one or more cluster nodes.

- **The speed/streaming layer** manages fresher data to support any analytics or advanced analytics that request real/real-time data. Although they can also be stored in similar repositories to the batch data, the repositories used for streaming purposes are distributed data streaming solutions, such as Kafka or other types of queues. The data is stored for a short period of time (usually up to one month).

  - **Fault tolerance**: Due to the storage infrastructure, it also can maintain proper operation despite failures.

  - **Back-feed between operational and analytics**: The integration between both layers is addressed by creating views for batch and speed data, enabling the data users and consumers to consume the data within the serving layer. Here, you can see the evolution of the ODS, which addresses storage needs in a more timely and scalable manner.

- **Data retention definitions**: Already mapped in the traditional data architecture (a.k.a. BI architecture), the data retention rationale was leveraged for the big data architecture. Besides on-premises solutions, cloud services have evolved and created solutions such as AWS Glacier, which was created to store cold data. The approach remains the same: create clear policies and layers for retaining or archiving data based on its significance and utility.

- **Specific modeling and storage definitions**

  - **Data modeling techniques**: Besides the technologies to be used as repositories for specific use cases, such as new data warehouses, distributed file systems, or No-SQL databases, new modeling techniques were also used. Adopting data vault modeling became another option besides the dimensional modeling techniques suggested by Inmon and Kimball.

  - **Metadata for data governance**: Besides the necessary work for analytics, metadata management was created for data governance, as stated by DAMA in 2009.

- **The necessity of a common data model**: It was necessary to move from the data-fragmented generation and consumption that occurred at the beginning of the big data era to a more unified data analysis within the big data platform. Although marked in the history of inventions before 2010, it lasted until 2019. By the way, it is still a work in progress for many companies. Anyway, this movement was fundamental to solving many issues, such as removing the delay to understand the data and make better decisions, reducing silos, redundancies, inconsistencies, and so on.

- **Flexibility for scalability**: Big data architecture also solved the issues related to scalability with the cloud infrastructures, as it's possible to scale the infrastructure to execute the services as soon as necessary, depending on the business requirements. This allows for on-demand flexibility and scalability.

- **Flexibility for cost management and reduction:**

  - **Horizontal infrastructure instead of vertical:**
    The costs of commodities decreased significantly
    compared to the past. On the contrary, the
    individual servers have a high capacity. For this
    reason, this strategy created for big data solved
    the necessity of powerful machine resources to
    support the exponential growth of data. Thanks to
    this, it was possible to have the birth and growth
    of many startups in the 2010 decade. The cloud
    infrastructure has become so popular due to its
    facilities that it's the most used approach in the
    market nowadays.

  - **Management of data storage by temperature:**
    With the retention of data services, new possibilities
    for data cost management increase the opportunity
    to reduce costs when policies are defined to
    manage the storage of the data accordingly. For
    instance, archiving less frequently accessed data
    while retaining easy access to critical information
    can help reduce storage costs.

# Continuing Challenges for Big Data Architecture

- **Unclear division of layers for the common data
  model architecture:** Although there are architectures
  that propose divisions of data layers for a common data
  model architecture, it's not clear how to apply it in a
  fashion that fits with the data mesh proposed of having
  data managed by the business domains owners.

- **Repository to integrate batch and online data**: Integrating batch and online data is still a concern. Although there are architectures to handle the way to work with each of them, it does not solve the necessity of integrating and maintaining both repositories, which are used for different purposes.

- **Data growth and high costs**: The cost of storing large amounts of data is still a concern. Although there is much more flexibility than in the past for storage cost management, this topic is still a challenge, especially if you think about the evolution of AI, which requires more data and, consequently, more storage.

- **Tool selection for big data storage**: Choosing the right tool for storage can be complex. Poor decisions made can lead to wasted effort, time, and resources.

- **Data security**: Big data repositories are attractive targets for hackers. For this reason, it is fundamental to continuously monitor strong security measures.

# Computing Limitations

The following have existed since the creation of big data architecture.

- **Ability to process comprehensive data storage**: The evolution of technologies and generation of data solved processing semi-structured and unstructured data. Now, technologies like Map Reduce were fundamentally created for Hadoop (influencing the creation of other solutions, such as Spark) to solve this type of issue. Besides, data warehouses also have functionalities that enable the computing data (ingestion and transformation) for semi-structured data, such as Redshift Spectrum, Google Big Query, and Snowflake.

- **Frameworks improving and reducing the necessity of using many technologies**: As soon as big data architecture evolved, it demanded more and different uses of technologies. A good example is Spark, which consolidated many functionalities within itself and removed the necessity of using many technologies. Besides improving the pace of developments, it also increased the velocity of computing data compared to Map Reduce.

- **Scalable infrastructure to compute massive data volume**: The big data strategy of using horizontal infrastructure solved the necessity of using powerful machine resources to compute exponential volumes of data. It started with the Hadoop Map reduction and evolved with the creation of Spark, which processes data in a distributed fashion.

- **Faster and more efficient processing, specific layers, and process definitions**: Latency of data for online usage by operational/analytics services. The Lambda and Kappa architectures solved the necessity of generating fresher data for analytics and advanced analytics. The Lambda architecture addresses the two main scenarios: batch and streaming data. This approach ensured fault tolerance, throughput, and latency.

  - **Fault tolerance**: This capability is fundamental for high availability for critical environments. The distributed data processing approach enables proper operation despite failures in one or more cluster nodes.

171

- **Throughput**: It's related to the volume of data being trafficked from its source until its target. This capability is fundamental for big data and is being addressed for batch and streaming with the Lambda and Kappa architectures and technologies implemented for each specific purpose.

- **Latency** is related to the delay between one event's occurrence and the data's availability for use. This capability is especially relevant to the streaming data approaches that were addressed with the Lambda and Kappa architectures.

- **On-demand flexibility for scalability**: Big data architecture also solved the issues related to scalability with the cloud infrastructures, as it's possible to scale the infrastructure to scale capacities to process big amounts of data that may increase in some seasonal periods. It is helpful for big data architectures and data warehouses that use MPP and columnar memory-based storage.

- **Flexibility for cost management and reduction**:

  - **Decouple of storage and computing**: With decoupled storage and computing, it's possible to define different machine configurations to attend to the demands of the business, bringing the possibility of strategically planning the Capex[43].

---

[43] Capex (short for capital expenditure) is related to the amount a company spends to acquire, maintain or improve its fixed assets. It's relevant because it contributes to the business's long-term growth, increasing productivity, and profits.

- **Horizontal infrastructure instead of vertical**:
  On-demand scalability of infrastructure enabled
  the reduction of costs by hiring just the necessary
  capabilities for the business, including seasonal
  scenarios requesting more capabilities. Very
  used by startups and become the most used in
  the market.

- **Management of data storage by temperature**:
  Once the cold data is moved to the appropriate
  storage, it won't be computed as warm or hot data,
  directly reducing cost management. That's one of
  the reasons for this policy.

- **Incremental data ingestion and processing**: Delta
  Lake features implement incremental data ingestion
  and data processing. That way, it solves the deficiency
  by requiring the full data processing for the whole data
  lakehouse.

# Continuing Challenges for Big Data Architecture

- **Unclear division of layers for the common data
  model architecture**: Although there are architectures
  that propose divisions of data layers for a common
  data model architecture, it's not clear how to apply the
  computing responsibilities in a fashion that fits with
  the data mesh proposed of having data managed by the
  business domains owners.

- **Necessary to learn and adapt the big data
  architecture for incremental ETL/ELT**: Although
  there are the Delta Lake features now to implement

incremental data ingestion and processing, except for the startups, every company that already has its own analytical environment needs to adapt their existing data pipelines to fit the incremental data loading functionalities: for ingestion and processing.

- **Lack of efficiency in generating datasets for analytics**: The creation of datasets that are compliant with the data governance policies once defined (and they should be to enable efficient data management).

- **Maintenance due to duplicated coding for batch and streaming data**: The maintenance of big data architecture is a concern for the business. The Lambda architecture leads to duplicated computation logic and the complexity of managing the architecture for both paths. Although the implementation is easier due to the logic being similar and can eventually be leveraged, it contains two systems requiring different paths for coding, one for batch and another for the speed layer. The Kappa architecture makes maintenance more complex and, eventually, removes key characteristics and goals of this approach.

- **Data growth and high costs**: The cost of computing large amounts of data is still a concern. Although there is much more flexibility than before for computing cost management, this topic remains a challenge, especially with the evolution of AI, which requires more data and, consequently, more capabilities to process the data.

# Data Management and Data Governance

The following have existed since the creation of big data architecture.

- **Data architecture design for data management**: with the new technologies, it's possible to scale the storage and computing necessary for big data. For easier management, there are proposed architectures to handle batch data related to the BI architecture and streaming data.

- **Self-service for data users and consumers**: With big data architecture, it is possible to enable self-service for data users and consumers. It started with the data lake and keeps evolving with the data mesh concepts.

- **The necessity of a common data model**: the creation of the data lakehouse and the medallion data architecture improved the consumption of data by creating more trustfulness for the generation of new business through traditional BI responsibilities such as regulatory reports, business monitoring, and also innovations generated through business processes such as experimentations and ML models.

- **Data governance policies: data quality and consistency**: The data governance definitions on DAMA enabled organizations to create their own policies based on standards and best practices definitions. Besides, data modeling and ETL/ELT techniques support the implementation of these policies for confidence in the data.

# Continuing Challenges for Big Data Architecture

- **Enterprise data culture and sponsorship**

  - **Culture**: one of the main challenges that remain, and now with great impact, is related to the necessity of a data analytics culture instead of just an analytics culture. While the analytics ad hoc results are the focus, we keep the fundamental issues that directly impact the AI evolution.

  - **Sponsorship**: The sponsorship for big data architecture must be the same as the interests related to AI, and ideally, because one is the foundation for the other. However, it's still a challenge in marketing because many leaders do not have clarity about these impacts.

- **The necessity of even more strategic data analytics leadership**: In the same way there is a challenge related to sponsorship for data, there are many data leaders (regardless of their title or specialization) that are not able to argue the business impact of the issues and create the adequate connection of the organizations needs and goals. For this reason, strong strategy vision, product skills, and technical knowledge are critical for data management and governance in the AI era.

- **Source of truth data (golden data)**: Although there are very good architectures and architecture strategies, the market still suffers from a lack of clarity and confidence in data accuracy. It mainly happens due to siloed data that can diverge in structure, quality, and standards. Even with the data mesh concepts, the practical approach of splitting and managing the data appropriately remains a big challenge for the market.

- **Clarity about roles and cross-functional collaboration**: Related to the golden data issue, the same occurs for this aspect. There's no clarity about roles and how to apply cross-functional collaboration to ensure data is generated as necessary to enable insights translated into actionable business strategies as organizations expect.

- **Professional legal compliance**: With big data and in the AI era, focusing efforts on legal compliance has become vital. After all, the laws can differ and evolve by country, and while data and AI keep evolving, the laws are also up to change and evolve.

- **Data governance**: One of the key points to ensure legal compliance and data management is data governance. The same applies here: while data and AI keep evolving, the definitions of data governance are also up to change and evolve. Actually, it must be. Everything mentioned by DAMA remains an important guideline. However, some special challenges must be considered: organization, cataloging, security, data quality, and consistency.

# 3.4. Summary

Let's recap the main takeaways of this chapter.

- The main business problems that incentivized the creation of the big analytics environment were (and still are) the necessity to address the following.

  - Even greater necessity to use and manage a big volume of data than after the 2000s with huge amounts and types of data

  - The necessity of intelligent and faster usage of data due to strong concurrences in the market regarding globalization

  - Ability to use data in a timely manner by not just having data for analytics purposes but also using it in operational activities as key features for products (e.g., recommendation of products and services)

  - Analytics data in operational services with appropriate latency for the business needs

  - Necessity of all types of data for discovery and analytics, including audio, videos, photos

  - Necessity of full data for analytics and advanced analytics: the historical and most recent data available for monitoring some operation in real-time, for instance

  - Ability to integrate operational solutions with analytics solutions

  - Criteria to manage the data life cycle and manage the costs appropriately

- Big data emerged from the necessity of evolving the capacity of traditional analytical architecture scalability. It comprises at least eight characteristics: volume, variety, velocity, value, veracity, visualization, viscosity, and validity.

- A data lake is a big data repository that contains structured, semi-structured, and unstructured data. Besides, it is the environment for data users for discoveries and tests.

- Lambda and Kappa architectures are related to data architecture design for big data Environments.

  The Lambda architecture contains three main layers: batch, speed, and serving layers.

  While the Kappa architecture contains just two layers: streaming and serving layers.

- Medallion data architecture was defined to address the clarity of data usability. It comprises three layers: bronze, silver, and golden data.

- A data lakehouse is the data management architecture that combines the data warehouse and data lake's best features to enable any analytics and advanced analytics data.

- Delta Lake is a functionality the Databricks created with the Linux Foundation to enable ACID transactions within HDFS and equivalent solutions. It started as a commercial solution but has an open source project, currently supporting solutions from different vendors.

- DataOps is an acronym for data operations and is a methodology that leverages DevOps, agile, and lean manufacturing to manage data through automation, monitoring, and collaboration among professionals while focusing on efficiency improvement. It also englobes the data contract concept. Super important to produce data products and is an important part of the Golden Data Platform, discussed in Chapter 5.

- A data contract is a component definition of DataOps. It's composed of terms and conditions defined between producers and consumers of data, with automation to ensure the data is transmitted accurately and consistently.

- Shift-left, a.k.a. shift-left testing, is a software testing approach created by Larry Smith. DataOps has leveraged this approach for data quality monitoring.

- Challenges solved with big data architecture: besides the business issues, there were technical challenges that needed to be solved and were addressed with the big data architecture. Remember that these issues are also related to the AI challenges, as it completely needs big data to perform and exist in a scalable manner as the AI era aims.

  - Storage

    - Big volume of data and management of storage issues

    - Ability to handle a comprehensive data storage

    - Scalable storage infrastructure

- Persistency of data according to the business use case

- Data retention definitions

- Specific modeling and storage definitions for analytics and data management

- Necessity of a common data model

- On-demand flexibility for scalability

- Flexibility for cost management and reduction

- Computing

  - Ability to process a comprehensive data storage

  - Frameworks improving and reducing the necessity of using many technologies

  - Scalable infrastructure to compute massive data volume

  - Faster and efficient processing, specific layers, and process definitions

  - On-demand flexibility for scalability

  - Flexibility for cost management and reduction

  - Incremental data ingestion and processing

- Data governance

  - Data architecture design for data management

  - Self-service for data users and consumers

  - Necessity of a common data model

  - Data governance policies: data quality and consistency

- Challenges remain even after big data architecture: Although many issues were solved, others emerged and increased in the technical aspects, such as the following.

    - Storage

        - Unclear division of layers for the common data model architecture

        - Repository to integrate batch and online data

        - Data growth and high costs

        - Tool selection for big data storage

        - Data security

    - Computing

        - Unclear division of layers for the common data model architecture

        - Necessary to learn and adapt the big data architecture for incremental ETL

        - Lack of efficiency in generating datasets for analytics

        - Maintenance due to duplicated coding for batch and streaming data

        - Data growth and high costs

    - Data governance

        - Enterprise data culture and sponsorship necessity

        - Necessity of even more strategic data analytics leadership

- Lack of source of the truth data (golden data)

- Clarity about roles and cross-functional collaboration

- Legal compliance

- Data governance definitions and compliance

## Summary

In this chapter we talked about important concepts regarding big data architecture, the issues that generated it, and the challenges and limitations. Next, let's consolidate the knowledge you have gained by comparing the two main architectures during the journey.

This knowledge is relevant as the foundation for the second part of this book, where we'll talk about the proposal of this book as a path to manage data as a product in practice.

# CHAPTER 4

# Consolidating the Analytics Knowledge Journey

You have probably identified many similarities among the architectures and definitions you read in this book, right? Great! That's the idea!

Instead of simply considering new solutions as the source of truth and the correct approach, consider all possibilities. To design the best solution for a specific problem you face, you must have two main skills: solid background and critical thinking.

The solid background you develop through real-world experience, curiosity, and as you have been creating since you started reading this book: studying. Remember, it's a journey. For critical thinking, my suggestion is for you to use your curiosity in your favor by asking questions and analyzing the information you have.

This chapter connects the issues and solutions implemented during our analytics journey thus far. You'll learn why it is important to have these fundamentals knowledge as ground to manage a product of data, regardless what it is: data product or data as a product. Most importantly, I want to be able to move forward to the next chapter, which explains the Golden Data Platform design.

© Jessika Milhomem 2025
J. Milhomem, *Data Product Management in the AI Age*,
https://doi.org/10.1007/979-8-8688-1315-3_4

# 4.1. Data Evolution Overview

To start the analyses and comparisons, let's first recap our analytics journey thus far.

## Relational Databases for Enterprise Applications

In the past, it was a huge manual effort to store data. From a bunch of challenges, studies, and work emerged the innovative solution to address it: the relational database created by Edgar F. Codd.

Since it was created, the organizations started to implement enterprise applications to automate solutions. Seeing the potential of the databases, of course, the organizations became interested in using this revolutionary innovation as a source of knowledge.

They wanted to study the behavior of everything relevant to the business and started analytics. One of the most relevant data for an organization is the same as ever: customer behavior.

Based on this need, database marketing, introduced by Robert Shaw, became one of the most important tools for marketing professionals. It was fundamental to perform customer relationship management (CRM). Of course, statisticians were also using the repository to perform advanced analytics, but on a much lower scale than there is nowadays with the renamed (and evolved, with new algorithms) data science techniques.

Data is important, especially now that globalization is affecting it.

## Data Warehouse Ecosystem

Under the watchful eye of this trend, Bill Inmon introduced the data warehouse with a revolutionary vision: implement a data architecture that has all the analytical data in a centralized repository.

This revolutionary approach didn't address the analytics data needs of just one business area but all of them in one unique architecture. It wouldn't be the ideal environment just for performing business intelligence (BI) but also for advanced analytics through statistics.

Yet, alert to the market's interests and needs, he identified the necessity of having analytical data available for applications to be used in an online mode, such as CRM. Then, he introduced a new concept: the operational data store (ODS), a repository with analytical data used for transactional operations to interface with a data warehouse.

Although the architecture concept is still great, the time required to implement a complete data warehouse supporting all the areas of an organization takes a long time to implement; typically, more than a year and without the possibility to take advantage of it before its full conclusion.

Then, Kimball introduces a new paradigm for a data warehouse. The fundamental values are leveraged from Inmon's vision, but the architecture design and modeling are different. It's modular. It's composed of business subjects (a.k.a. data marts). The main advantage here is that it's possible to take advantage of the data sooner because it's not necessary to implement a whole data warehouse but one specific business subject after another. For example, instead of the marketing team waiting more than a year, they can take advantage of the data in a few months.

# Big Data Ecosystem

Globalization is the reality, and the internet and data are part of humanity's life. Technologies are evolving and becoming more accessible to the population.

Data is now generated massively. Big volumes, with different formats and types of data, from different environments and different subjects. Not just professional information. Data about everything.

Companies have a huge opportunity to create new businesses by using this vast and diverse volume of data for analytics and advanced analytics.

Therefore, those who better leverage their (own and outside) data to create and execute their strategy by using a scalable architecture win!

For this reason, the ability to handle it is the focus of many companies.

Google innovates and creates its own solutions, but it also collaborates with the community by sharing through papers the fundamental concepts generated by their discoveries. These materials are supposely also used by Yahoo to develop Hadoop, the distributed framework to handle this new era of data. They also share it with the community, enabling it to be open source.

With lots of contributors (from companies or individual contributors) working on these projects, new solutions are created to handle data in different aspects: discoverability, computing, and others. Overall, the tools are for the same purpose but with different pieces for each step of the data, especially for data wrangling.

In terms of business, data users now have the ability to get the data they need at any time and as they want. They do self-service. The reason for this is the concept of the data lake, which uses the schema-on-read approach. This means that data wrangling occurs when they want to use the data, not before that, by having it available just for usage.

It's also great for data science projects that can leverage these benefits.

However, with this new approach to the data lake, new issues (already known and solved in the BI environment) start to become painful: no one has clarity and confidence in the data because the data doesn't exist as a source of truth.

Thus emerges the medallion data architecture, which creates layers of data to classify and ensure visibility of the type accuracy. Together with it, the concept of the data lakehouse appears. The market understood that it was important to consider some traditional architecture and leverage its features to solve known problems on a different scale.

In terms of technical aspects, many solutions to solve pieces and different types of problems create a huge maintenance effort and complexity. Around this issue, Spark emerges from a project by researchers at the University of California, Berkeley. This project strongly simplifies computing processes, as it addresses different types of projects while supporting and guaranteeing compatibility with a variety of programming languages.

Another issue is in evidence: data latency.

The market needs near/real-time data besides batch data. New architecture designs emerge: Lambda and Kappa to address it in the best approach possible and with some similarity with the traditional one, although not directly related.

The time to compute the data still remains a problem for the batch because, contrary to business intelligence and ETL/ELT approaches, the big data didn't enable incremental data processing but just full data loading. The results are high costs and a long period to compute the data for a data lakehouse.

Then, the most recent innovation happened for big data: the delta lake was created to enable incrementalization and push organizations to start using it.

# Artificial Intelligence Evolving

In the big data era, it is possible to implement, test, and evolve many advanced analytics solutions. Besides using statistics, it's possible to use mathematical and other sciences to evolve the algorithms used for machine learning, as mentioned in Chapter 1.

Although in the (maybe near) future, we'll have efficient generative AI models, or the artificial general intelligence, autonomous and smart enough to distinguish not accurate from accurate data or even to use any kind of data to create accurate data for predictions, we are not there yet. We need to create and ensure accurate data to train the models.

It is not a pain just for BI projects in the big data environment but also for AI/advanced analytics projects. After all, how can models be created to predict scenarios if the actual data used for this purpose is not accurate? Clearly, the potential for inaccurate predictions is high without golden data.

Defining the data lakehouse and medallion architectures helped with that but didn't fully solve it.

## Data Mesh

Besides all these challenges, another one that continues to be latent is related to ownership, generation, and management. With so much data, it's not clear how to organize teams, what the scopes of generation and consumption are, and how to ensure minimum governance.

Sitting on these issues, Zhamak Dehghani proposed the data mesh approach by connecting four main approaches to it: domain-oriented ownership, data as a product, self-serve data platform, and federated computational governance. This approach is the most recent one in the market and is being discovered and tested.

Next, let's evaluate each one of the components in the architectures proposed so far.

# 4.2. Traditional BI and Big Data Architectures

Let's evaluate some similarities here in terms of problems to be solved, context, or design. I know that you have identified at least one of them.

The main goal of both architectures is to be one centralized environment in which to perform analytics and advanced analytics initiatives.

We know the definition of data warehouse architecture may vary when compared to Inmon's or Kimball's vision, but regardless of its definition, a data warehouse contains values that are similar to big data. Considering the order of history, big data has similarities with data warehousing.

# Similarities

- **Both architectures do not use the transactional repositories for analytics** responsibilities, as besides they have different main purposes, they also have distributed information due to their operations that, though about similar business domains, are related to different operations.

- **Both architectures are commonly referred to as ecosystems**, not just one repository or mechanism. They have different components as part of them, but they have similar purposes: ingest raw data, process it, and make it available for analytics and advanced analytics.

- **Both contain data from all business domains**, not just one specific business domain, although they can be evaluated from different perspectives and formats.

- **Both are time-variant and generate value**, which means that both contain the historical and most current data available for analysis and pursue the capacity to extract and generate value with them, regardless of the period.

- **Both are integrated and handle a large volume of data** as they integrate data from all sources to ensure the veracity of the data, considering all possible sources. With DW 2.0, it has the ability to handle semi-structured and structured data as big data.

- **Both contain enterprise data and aim to ensure the veracity of information**, not disinformation, due to the new inclusion of values related to it for big data.

- **Both architectures have ML models trained and executed in a batch mode**, usually using the data warehouse, which is responsible for storing the results for analytics evaluation or other possibilities for the business. In a data warehouse, it is usually used for marketing campaigns and CRM.

# Differences

- **Subject-oriented**: While a Data warehouse is modeled by domain, big data has this characteristic only when it's modeled as a data lakehouse. Besides, the data mesh is bringing this concept to big data architecture again through domain-oriented ownership.

- **Variety and big volume**: Although both are integrated, just Big Data has the ability to ingest and handle unstructured data and, consequently, the volume associated with it.

- **Data governance reinforced**: Although relevant for a Data warehouse, as it was created for BI architecture initially, it became even more important and critical and not optional with the big data.

- **Focus on accurate data**: The data warehouse was created to be the source of truth for analytics. For this reason, it is mainly implemented in ETL mode.

  Big data was initially designed to support self-service with the data lake, regardless of the accuracy of the data. Recently, it has incorporated these characteristics with the data lakehouse concept. Thus, one of the differences you can see in both ecosystems is that one is fundamentally designed to serve accurate data, while the other one can deliver it, depending on its design.

By the way, one of the reasons for additional V values included in the big data definition, such as veracity and validity, is the importance of the data for fundamental business processes.

- **Non-volatile**: A similar behavior occurs with the volatility when comparing a data warehouse and the big data. While it's a fundamental value for a data warehouse as part of its definition, the big data may deliver it if it is designed as a data lakehouse, and ideally using the delta lake features. As you can see, it is a recent feature that was created. Before that, its behavior would have been possible if it had been simulated with different approaches, but again, it would not have been a mandatory definition of big data.

- **Online execution of ML models**: With the near/real-time capabilities, the execution of ML models is possible with the big data architecture.

**Similar Consumers**

- **For analytics**: Business intelligence analysts perform discoveries within the repositories, and business users consume the built reports and dashboards. Eventually, applications consume data from the ODS/repositories from the speed/serving layer.

- **For advanced analytics**: Statisticians are responsible for evaluating the data and creating statistical models. Compared with the current scenario, they are the data scientists who design the ML models. Also, the business users will consume the final predictions of the ML models that have been implemented.

**Different Consumers**

- **For advanced analytics**: Machine learning engineers implement it in production, both online and in batches. Applications consuming the ML models in real-time.

**Similar Developer Team Members**

- **For data modeling and ETL**: For both architectures, it's necessary to have professionals to work with data modeling and ETL/ELT. In a data warehouse architecture, there are DW/ETL architects who are currently analytics engineers or data engineers. Similar to the ML model design work, this work evolved by including new technologies and approaches in the scope.

- **For analytics**: Business intelligence analysts are responsible for making discoveries and use advanced analytics solutions to daily support the business decisions. They are also called business analysts or business analytics analysts.

- **For advanced analytics**: Statisticians or data scientists

**Different Developer Team Members**

- **For infrastructure**: Obviously, we need professionals not just to design but also to continue the architecture. In a data warehouse environment, there are database administrators who aggregate abilities and now are the database reliability engineers. For big data, due to the variety of components, as discussed, there are data platform admins, which can also be performed by software reliability engineers or database reliability engineers.

- **For advanced analytics**: To enable the usage of AI/ML models integrated with the services, not just in batch, but also in online mode in big data, a new role emerges: the machine learning engineer.

In short, you can see that a data warehouse ecosystem evolved into big data, which also has been evolving to consolidate even more functionalities due to the expansion of data characteristics and the emergence of new needs. It's also continuously adding traditional ones. Table 4-1 summarizes and compares a data warehouse and big data.

***Table 4-1.*** *Data Warehouse vs. Big Data*

|  | SIMILARITIES | DIFFERENCES |
|---|---|---|
| **CHARACTERISTICS** | • Architectures do not use transactional repositories for analytics. <br> • Architectures are commonly referred to as *ecosystems*. <br> • They contain data from all business domains. <br> • They are time-variant and generate value. <br> • They are integrated and handle a large volume of data. <br> • They contain enterprise data and aim to ensure veracity. <br> • They execute ML models in batches. | • A data warehouse is designed to be subject-oriented and non-volatile and serve accurate data. <br> • Although some functionalities for data warehouse, only big data is capable of working with a large variety and massive volume. <br> • Although designed for data warehouses, data governance is mandatory for big data. <br> • Only big data enables the online execution of ML models. |

(*continued*)

*Table 4-1.* (*continued*)

|  | SIMILARITIES | DIFFERENCES |
|---|---|---|
| **CUSTOMERS** | **For analytics**<br>• Business Intelligence analysts / Business analysts / Business Analytics analysts<br>• Applications<br>**For advanced analytics**<br>• Statisticians / Data Scientists<br>• Business users | **For advanced analytics**<br>• Big data: Machine Learning Engineers<br>• Big data: Applications |
| **DEVELOPER TEAM** | **For data modeling and ETL**<br>• Data warehouse and ETL architects / Analytics Engineers / Data Engineers<br>**For analytics**<br>• Business intelligence analysts / Business Analysts / Business Analytics Analysts<br>**For advanced analytics**<br>• Statisticians / Data Scientists | **For infrastructure**<br>• BI: Database Administrators / Database Reliability Engineers<br>• Big data: Data platform admins / Database Reliability Engineers / Software Reliability Engineers<br>**For advanced analytics**<br>• Big data: Machine Learning Engineers |

**Note**   It's relevant to mention that some companies merge the responsibilities of the Machine Learning Engineer with the Data Engineer or Analytics Engineer roles. It may be a tendency in future.

# Staging and Data Lake

The main goal of both architecture components is to be one centralized repository with the as-is data from the sources (raw data) for analytics purposes.

## Similarities

- **Both architecture components handle raw data.** The staging and the data lake repositories receive and store the ingested data from the sources.

- **Both architecture components are used for discoverability.** Although with different intensities and performed by different profiles, the repositories are used for discovery. Professionals performing the ETL processes explore and analyze the data and make relevant information available in a data warehouse. It is not only the professionals performing ETL; business professionals also do explorations and experimentations in the data lake.

- **Both architecture components are used as sources for data warehouse and data lakehouse environments.** They coordinate data integration from different sources and then make the data available. The synchrony of the raw data is super important to make sure the data wrangling considers comparable data to generate accurate insights.

# Differences

- **Variety of data**: Staging supports structured and semi-structured data, while the data lake supports all (structured, semi-structured, and unstructured data).

- **Unlimited business exploration and experimentation**: A data lake has unlimited business exploration and experimentation access. Here, the behavior is more similar to the traditional data warehouse concept (especially suggested by Inmon) regarding access and usability than staging.

- **Accessible by business users**: Only the data lake enables business users to use it through queries or presentation layers.

## Similar Consumers

- **For advanced analytics**: Statisticians and data scientists design the ML models, and the business users consume the final predictions of the ML models implemented.

- **For data modeling and ETL**: Data warehouse/ETL architects, analytics engineers, or data engineers are responsible for ingesting, treating the data, and loading it into a data warehouse or data lakehouse.

## Different Consumers

- **For analytics**: Business intelligence analysts are responsible for discoveries within the repositories and creating their datasets.

**Similar Developer Team Members**

- **For data modeling and ELT**: Data warehouse/ETL architects, analytics engineers, or data engineers are responsible for implementing the ingestion pipelines to load the staging or the data lake.

- **For advanced analytics**: Statisticians or data scientists use for some exploration and when using raw data for models.

**Different Developer Team Members**

- **For infrastructure**: It's the same result as comparing a data warehouse and big data.

- **Ingestion of data**: In big data, there are still scenarios where data platform admins also automate some types of ingestions.

- **For analytics**: In the staging area, initially, only data warehouse/ETL architects and data scientists had access and were able and responsible for discovering data. However, now, with the big data context, it's open and in the data lake. For the data lake environment, there are analytics engineers/data engineers and business intelligence/business analytics analysts.

- **For advanced analytics**: Statisticians, data scientists, and machine learning engineers may implement ML models by using raw data.

In short, you can see that the staging evolved into a data lake, which has also been evolving to consolidate even more functionalities due to the expansion of data characteristics and the emergence of new needs. Table 4-2 summarizes and compares staging and a data lake.

Table 4-2. Staging vs. Data Lake

| | SIMILARITIES | DIFFERENCES |
|---|---|---|
| CHARACTERISTICS | • Architecture components handle raw data.<br>• Architecture components are used for discoverability.<br>• Architecture components are used as sources for data warehouse and data lakehouse environments. | • Variety of data<br>• For data lakes: Unlimited business exploration and experimentations<br>• For data lakes: Accessible by business users |
| CUSTOMERS | For advanced analytics<br>• Statisticians / Data Scientists<br>For data modeling and ETL<br>• Data warehouse and ETL architects, Analytics Engineers / Data Engineers to feed data warehouses or data lakehouses | For analytics<br>• Data lakes: BI analysts/ Business analysts/ BA Analysts |

(continued)

*Table 4-2.* (*continued*)

| | SIMILARITIES | DIFFERENCES |
|---|---|---|
| DEVELOPER TEAM | **For data modeling and ELT**<br>• Data warehouse and ETL architects / analytics engineers / data engineers<br>**For Advanced Analytics**<br>• Statisticians / data scientists | **For Infrastructure**<br>• Staging: Database administrators / database reliability engineers<br>• Data lake: Data platform admins / database reliability engineers / software reliability engineers<br>**Ingestion of data**<br>Data lake<br>  • Analytics engineers / data engineers<br>  • Data platform admins<br>**For analytics**<br>Staging<br>  • Data warehouse and ETL architects / analytics engineer / data engineer<br>  • Statisticians / data scientists<br>Data lake<br>  • Analytics engineers, data engineers / business intelligence analysts / business analysts<br>**For advanced analytics**<br>  • Statisticians / data scientists<br>  • Machine learning engineers |

# Data Warehouse and Data Lakehouse

The main goal of both architectures is to be one centralized stack with accurate data (a.k.a. golden data) for analytics and advanced analytics purposes.

## Similarities

- Both consume data from all sources available.

- Both require historical data for analytics besides the most updated data.

- Both are centralized architectures that produce and, as a repository store, make available accurate data for analytics purposes.

- Both handle subject domains (a.k.a. business domains) via data marts.

- Both must ensure data quality controls are in place to ensure the data is accurate and usable.

- Both are accessible for business and technical users to consume the data via querying or presentation/ serving layer.

- Both contain data that may be used for operational purposes, although they have different scalability.

- Both enable complete data management of the architecture, ensuring the governance for the whole stack.

- Both maintain silver data and make it available for consumption.

# Differences

- Only data lakehouses work with near/real-time data in a large volume.

- Although there are some cases in the community where silver data is used for data warehouses and golden data is used for data marts, in the majority of cases, the data lakehouse maintains the silver data and makes it available for consumption.

- Only data lakehouses contain any type of data (unstructured, semi-structured, and structured) and enable any analytics and advanced analytics.

- Only data lakehouses support applications with analytical data on a large scale.

### Similar Consumers

- **For analytics**: Business intelligence analysts perform discoveries within the repositories, and business users consume the built reports and dashboards.

- **For advanced analytics**: Statisticians and data scientists are responsible for designing the ML models.

### Different Consumers

- **For advanced analytics**: Machine learning engineers implement it in production for online or batch. Applications consuming the ML models in real-time. Business users consume the ML model predictions.

### Similar Developer Team Members

- **For data modeling and ETL**: Data warehouse and ETL architects, analytics engineers, data engineers

- **For analytics**: Business intelligence analysts / Business Analysts / Business Analytics analysts

- **For advanced analytics**: Statisticians and data scientists design the ML models.

### Different Developer Team Members

- **For infrastructure**: In a data warehouse—the database administrators / database reliability engineers; in a data lakehouse—the data platform admins / software reliability engineers / database reliability engineers

- **For advanced analytics**: Machine learning engineers put ML models into production (batch or online).

You can see that a data warehouse evolved into a data lakehouse, which has also been evolving to consolidate even more functionalities due to the expansion of data characteristics and the emergence of new needs. Table 4-3 summarizes and compares a data warehouse and a data lakehouse.

*Table 4-3. Data Warehouse vs. Data Lakehouse*

| CHARACTERISTICS | SIMILARITIES | DIFFERENCES |
|---|---|---|
| | • Consumes data from all sources available | • Data lakehouse: work with near, real-time data in large volume |
| | • Requires historical data for analytics besides the most updated data | • Data lakehouse: maintains the Silver data and makes it available for consumption |
| | • Centralized architectures that produce, and as a repository, store and make available accurate data for analytics purposes | • Data Lakehouse: contains any type of data (unstructured, semi-structured, and structured) and enables any type of analytics and advanced analytics |
| | • Handles with subject domains (a.k.a. business domains) via data marts | • Data lakehouse: supports applications with analytical data on an extraordinarily large scale |
| | • Ensures data quality controls to make sure the data is accurate for usability | |
| | • Accessible for business and technical users to consume the data via querying or presentation, serving layer | |
| | • Contains data that may be used for operational purposes, although in different scalability | |
| | • Enables complete data management of the architecture, ensuring the governance for the whole stack | |

*(continued)*

*Table 4-3.* (*continued*)

| | SIMILARITIES | DIFFERENCES |
|---|---|---|
| **CUSTOMERS** | **For analytics**<br>• Business intelligence analysts / Business Analysts / Business Analytics Analysts<br>• Applications<br>**For advanced analytics**<br>• Statisticians / data scientists<br>• Business users<br>• Applications | **For advanced analytics**<br>• Data lakehouse: Machine learning engineers<br>• Data lakehouse: Applications |
| **DEVELOPER TEAM** | **For data modeling and ETL**<br>• Data warehouse and ETL architects / analytics engineer / data engineer<br>**For analytics**<br>• Business intelligence analysts / Business Analysts / Business Analytics Analysts<br>**For advanced analytics**<br>• Statisticians / data scientists | **For infrastructure**<br>• Data warehouse: database administrators / database reliability engineers<br>• Data lakehouse: data platform admins / database reliability engineers / software reliability engineers<br>**For advanced analytics**<br>• Data lakehouse: machine learning engineer |

# ODS and Speed/Streaming Layer

The main goal of both architecture components is to be one layer to compute data for operational needs and, as a repository, be the interface with the transactional environment by serving accurate data (a.k.a. golden data) for analytics and advanced analytics.

## Similarities

- Both components perform as interfaces with the transactional environment by serving accurate data from for analytics and advanced analytics.

- Both components enable treatments inside its layer and storage of its information.

## Differences

- The ODS can become another source for a data warehouse, while the source from the operational environment is usually the applications.

- In the streaming layer, we can now work with more capacity and scalability due to new technologies emerging, such as No-SQL databases, and distributed data streaming solutions, such as Kafka.

### Similar Consumers

- **Transactions**: Analytical applications, such as CRM

- **For analytics:** It can be used by ingesting data from the ODS or consulting the speed views. In that scenario, the data can be used by business users.

- **For advanced analytics**: It can also be consumed by statisticians and data scientists for eventual data models.

**Different Consumers**

- N/A

**Similar Developer Team Members**

- **For data modeling and ETL**: Data warehouse and ETL architects/analytics engineers/data engineers

- **For analytics**: Business intelligence analysts / Business Analysts / Business Analytics Analysts

- **For advanced analytics**: Statisticians and data scientists design ML models

- **Transactional applications**: Software engineers

**Different Developer Team Members**

- **For infrastructure**: For ODS—database administrators and database reliability engineers; for streaming layer—data platform admins, software reliability engineers, database reliability engineers

- **For advanced analytics**: Machine learning engineers put ML models into production in online mode.

You can see that ODS evolved into the speed/streaming layer, which has also been evolving to consolidate even more functionalities due to the expansion of data characteristics and the emergence of new needs. Table 4-4 compares ODS and a speed/streaming layer.

*Table 4-4.* *ODS vs. Speed/Streaming Layer*

|  | SIMILARITIES | DIFFERENCES |
|---|---|---|
| **CHARACTERISTICS** | • Perform as an interface with the transactional environment by serving accurate data for analytics and advanced analytics<br>• Enable treatments inside its layer and store its information | • ODS: Can become another source for a data warehouse<br>• Streaming layer: Works with new technologies such as No-SQL databases and distributed data streaming solutions |
| **CUSTOMERS** | **Transactions**<br>• Analytical applications, such as CRM<br>**For analytics**<br>• Business intelligence analysts / Business Analysts / Business Analytics Analysts<br>**For advanced analytics**<br>• Statisticians / data scientists | N, A |
| **DEVELOPER TEAM** | **For data modeling and ETL**<br>• Data warehouse and ETL architects / analytics engineers / data engineers<br>**Transactional application**<br>• Software engineers<br>**For analytics**<br>• Business intelligence analysts / Business Analysts / Business Analytics Analysts<br>**For advanced analytics**<br>• Statisticians / data scientists | **For infrastructure**<br>• ODS: Database administrators / database reliability engineers<br>• Streaming: Data platform admins / database reliability engineers / software reliability engineers<br>**For advanced analytics**<br>• Streaming: Machine learning engineers |

# Data Marts in BI and Big Data Architectures

In general, the main goal of a data mart is the same regardless of the architecture used (see Chapter 3).

Table 4-5 summarizes and compares data marts in business intelligence and big data architectures.

***Table 4-5.*** *Data Marts in BI and Big Data Architectures*

|  | SIMILARITIES | DIFFERENCES |
|---|---|---|
| **CUSTOMERS** | **For analytics**<br>• Business intelligence analysts / Business Analysts / Business Analytics Analysts<br>• Applications<br>**For advanced analytics**<br>• Statisticians / data scientists<br>• Business users | **For advanced analytics**<br>• Big data: Machine learning engineers |
| **DEVELOPER TEAM** | **For data modeling and ETL**<br>• Data warehouse and ETL architects / Analytics Engineers / Data Engineers<br>**For analytics**<br>• Business intelligence analysts / Business Analysts / Business Analytics Analysts<br>**For advanced analytics**<br>• Statisticians / Data Scientists | **For infrastructure**<br>• BI: Database administrators / database reliability engineers<br>• Big data: Data platform admins / database reliability engineers / software reliability engineers<br>**For advanced analytics**<br>• Big data: Machine learning engineers |

# 4.3. The Analytics Journey

Now that we consolidated the knowledge, and did the comparisons, let's discuss the learnings we can get from the journey we've been studying thus far. The following are some examples of issues.

- Organizations focusing on schema-on-read only and ignoring a data warehouse architecture and resolutions consequently caused the companies to lose fundamental capabilities generated by data warehouses.

- Due to the lack of incremental ingestion and loading for batch, long-period computing requires a massive amount of data with a full ingestion and loading approach.

- Postponing or neglecting data governance is fundamental to efficient data management.

- Difficulty in composing teams due to the focus on skills related to technologies only instead of considering experience and knowledge of contexts and processes. Especially when comparing the business processes and necessities.

This is a simple and short list, but you can clearly map some important learnings!

# Learn from Issues and Challenges

Everyone must learn about the issues and challenges—your direct customers, your team, everybody!

The problem is the source of everything. The more you know about it, the better your discoverability work and solution design will be.

## Don't Ignore Experiences

Instead of simply considering new technologies and innovations as a silver bullet, create a solid background.

Of course, learning the role history of the specialty/business context you are working with is a key part of it.

During the journey, also learn about the existing solutions for each issue and problem that already exists.

## Embrace Innovation: Discover and Create New Solutions

We must also keep learning from innovations generated by so many talented people worldwide.

## Maintain the Cycle

It is similar to the circle of life, as said in *The Lion King* movie: It never ends! We need to continuously follow it in our routine. Learning, testing, and practicing are part of the work.

If you are not curious and persistent, I'd say this work is not for you because you must create a repertory and think critically about it. But if you enjoy it, you may succeed!

After reading this book, following the tips, and merging your skills and expertise, I'm sure you will be able to design a good strategy to help you mitigate many risks and issues and save time for yourself, your team, and your company!

# 4.4. Summary

Let's recap the main topics of this chapter.

We have similarities between the BI and Big Data architectures that are related to:

- BI and big data architectures

- Staging and data lakes

- Data warehouses and data lakehouses

- ODS and speed/streaming layers

- Data marts in BI and Big Data Architectures

The following are lessons you can learn from the journey.

- Learn the issues and challenges faced.

- Don't ignore lived experiences/learn from history.

- Learn the existing solutions.

- Embrace innovation: Discover and create new solutions.

- Maintain the cycle.

# Summary

Congratulations, you've just finished Part I of the book!

In this chapter we summarized the business intelligence and big data architectures and mapped lessons for data product management.

Next, let's move on to Part II of the book.

The next chapter talks about the Golden Data Platform, a strategy that I designed to implement and manage Data as a Product. We'll cover since the explanation about it, until how to implement it.

# PART II

# Data Product Management in Practice

# CHAPTER 5

# The Golden Data Platform to Manage Data as a Product

Although there are several innovations to solve many of the problems discussed in this book, issues remain, and new ones appear.

The next chapters in this book are related to a practical approach to managing data-related products, regardless of the type.

Finding no other complete solution in the industry today for all the problems discussed in this chapter, I have proposed a new approach that I call the Golden Data Platform. It is a solution to create and manage data as a product. This chapter explains what it is and discusses the problems it addresses, how it solves them, and the pre-requirements to design and implement it.

Let's learn how to manage data as a product through the Golden Data Platform design.

J. Milhomem, *Data Product Management in the AI Age*,
https://doi.org/10.1007/979-8-8688-1315-3_5

# 5.1. The Golden Data Platform Problems

There are new innovative concepts in the market related to business intelligence (BI) architecture. For instance, a data lakehouse is now the environment to forge accurate data. The data mesh also stimulates working with data as a product while focusing on specific domains.

Despite these very nice new solutions, there is still unclarity about how to solve the remaining issues.

- **Low business ownership and expertise for cross subjects**: Everything begins with the problem for the business. It's mainly necessary for accurate data, especially when discussing data as a product. Although the data mesh suggests the full ownership of the data by domain, it's unclear how to implement it in practice within the architecture and all the effort related to that.

  Even in a specific domain, there's a world of business contexts. Subdomains are inside it. So, the question remains: How can we implement and maintain it by ensuring business specifics? Even more complex: how do we implement and maintain it when there are cross-definitions and its usability?

- **Lack of modularity and governance**: When you think about a data lakehouse, some questions come up related to its modularity, such as how to implement the architecture while keeping flexibility to implement and use data in a modular manner?!

  The data mesh proposes that each team focuses on its own domain, which supports modularity ownership and responsibility, but it's up to the teams to define the approach. The only requirement is to ensure the federated guidelines for governance are compliant. Although the concept is great, it's still challenging to ensure an efficient modular solution while keeping an enterprise-level architecture and ensuring governance pragmatically.

- **Lack of standardizations**: The data lakehouse or data warehouse is responsible for ensuring accurate data. However, there are no guard rails defined to ensure it. Basically, it can be implemented in different formats or, regrettably, not be implemented at all. It can worsen when globalized solutions are implemented, even within a specific domain.

  Imagine the size of this issue when you consider it from an enterprise-level perspective.

- **Low efficiency for data factoring and management**: If there are no standards, you need to implement things from scratch every time. Consequently, the possibilities of automation are limited, and datasets need to be developed manually, which takes long time periods.

- **Lack of ownership visibility**: It's clear that ownership is a problem, and domain-oriented ownership of the data mesh's pillar is one important guideline for solving it. However, it's a challenge to put into practice this resolution. A clear necessity is the visibility of ownership. It's fundamental, together with the visibility of new dataset generation. It's even more relevant with business subjects used across different areas of the same business domain, and it's not necessarily clear in the metadata and in terms of the maintenance of everything and who the owner of it is.

- **Lack of efficient observability**: It's great that there are many tools to address observability and data hub solutions to consolidate them in one place. However, monitoring makes it more functional and efficient to have a specific and centralized place for it. Besides, the same standardization needs and contexts apply to it.

  A data hub is excellent for enabling the easy discovery of tools, not for observability as a work process.

- **Complex and maintenance not integrated**: The DAMA suggests data quality control implementation, and its guidelines are excellent guides for the type of issues. However, it's up to the teams how it'll be implemented. If it's implemented individually along the organization, it's not possible to visualize possible impacts suffered by other domains due to these impacts.

  Besides, there are similar issues related to the lack of standardizations, observability, and efficiency, but for this context. In other words, it is impossible to monitor anything without observability. Without standards, different processes and implementations create complex and unintegrated maintenance processes for the teams. Consequently, it generates even more inefficiency.

- **Lack of data governance integration**: Data governance is necessary, and these days, whoever works with data won't argue on the opposite side.

  However, everything for it depends on metadata. The challenge faced nowadays is related to the processes that are generated apart from the main business context. They are usually outside of everything, which creates additional effort and complexities for the professionals.

- **Inefficient data cost management**: Besides all the inefficiencies related to the execution of the creation of the datasets by itself, there's another critical topic related to that: the clarity of usage of the data for specific business processes. If we cannot map the usability of it, we cannot map the benefits of the data. Then, its generation becomes inaccurate and potentially a (huge) waste of money.

  Even if it occurs, it's still a problem that should be part of the data cost management work to address it appropriately.

# Golden Data Platform

I designed the Golden Data Platform to address all these issues.

## Platform Definition

According to Gartner,[1] a platform is a product that serves or enables other products or services. It ranges from a high-level platform to enable business models to a low-level platform that provides capabilities, other products, or services to deliver their own.

It usually serves consumers through APIs but may offer other mechanisms for consumption.

---

[1] Gartner.com, Glossary: (Platform Digital Business). Available at https://www.gartner.com/en/information-technology/glossary/platform-digital-business

To put any technical solution in place, some layers of architecture must be set up as a prerequisite in on-premise surroundings. For this reason, many clouds offer some platforms, from IaaS (infrastructure as a service), where the physical infrastructure is managed by the cloud companies, to SaaS (software as a service), where not just the infrastructure but also the service is managed by cloud companies.

The Golden Data Platform focuses on developing applications that manage the data (see Figure 5-1).

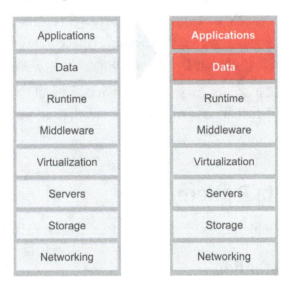

***Figure 5-1.*** *The Golden Data Platform focuses on data application and management*

# The Platform

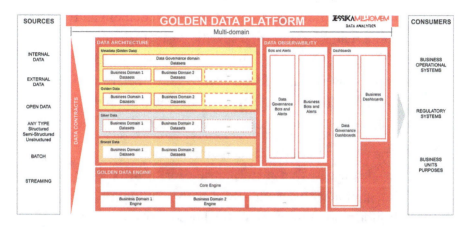

***Figure 5-2.*** *The Golden Data Platform*

The Golden Data Platform is a data product that generates and manages data as a product. It's business-driven, designed, and layered by domain, usability, and access control to deliver data as a product. It delivers data as a product with modular intelligence, solving and simplifying data management to create and maintain the ecosystem.

For this, the Golden Data Platform handles the data consumption and manages the data production for business intelligence and advanced analytics needs while ensuring data governance capabilities for complete ongoing maintenance (see Figure 5-2).

We'll talk about each one of the components of the architecture in the following sections, but let's summarize the most important components of it for general clarification.

- **Data contracts** are definitions (the contracts) used to ensure the integration between the data producer and the data consumers. These are files with standard definitions complying with all the agreed-upon requirements: schema definition, data quality, security and privacy, versioning, and so on.

223

- The **Golden Data Engine** is responsible for creating and managing data with standard best practices and defining requirements, considering each domain and subdomain's detailed and specific business definitions. This solution is modular by domain and subdomain and is fully managed (created and maintained by the owners of the domains). It is implemented upon the Itaipu data lake to generate the data lakehouse.

  The Golden Data Engine contains components inside the core that are responsible for ensuring integration with eventual third-party solutions or in-house solutions for data quality controls and other functionalities. If the functionalities don't exist, they are created as features inside the platform.

  Another important responsibility is to ensure the creation of the datasets inside the platform inside each layer of the data architecture.

  The Golden Data Engine is also composed of business domain engines responsible for managing the creation of the datasets, ensuring standardized definitions for the same domain. Thus, its implementation is modular and driven by the business domain.

- **Data architecture**: The data architecture follows standard definitions of domains and subdomains and is layered-based. It's fed with datasets created by the Golden Data Engine automatically.

  Each layer has security accesses, such as type of usage and private and public access.

- **Data**: The Golden Data Engine creates the datasets, specifically the business domain engines. They are created inside the data architecture and are products by themselves (data as a product), as they have business purposes, expected results, and customers and need to be maintained.

- **Data observability:** The data observability is composed of alerts and dashboards that are created upon the datasets created by the Golden Data Engine to enable faster monitoring, visibility, and analytics for business needs and data quality controls. Remember that all data products are integrated and managed in a modular way. In other words, by data domain. It can be created manually or automatically via the Golden Data Engine, depending on the data visualization tool's features, which are defined and used in the company's infrastructure.

With more clarity about each component of the Golden Data Platform, let's understand the creation of the golden data process inside the platform.

- For a specific domain, the data contracts are defined by the data services' producers and the data consumers. These contracts are standardized and generated automatically by the services and consumed by the data customers.

---

**Note**   The data consumers can also be data producers. In this context, they are data consumers of the service's data but are data producers for analytics.

These profiles can be or not part of the same team.

---

- The data domain owner must implement the Golden Data Engine's **core components** to ensure the implementation of domain-agnostic features, such as automation to **read schemas and data with agnostic definitions**.

- The data domain owner must implement the golden data business domain's **engine's features** to ensure the **consumption of the data, considering the contract definitions already considering the domain-specific definitions**.

- The data domain owner must implement the Golden Data Engine's **core components** to ensure the implementation of **domain-agnostic features** to **enable the creation of datasets in the appropriate domain and layer of the data architecture**.

- The data domain owner must implement the Golden Data Business Domain's engine's features to ensure data production, considering the domain-specific definitions.

- The Golden Data Engine **generates datasets** related to the business domain definitions **inside the data architecture** in the appropriate layers, considering the design of the solution.

- **Alerts** are **created automatically** by the Golden Data Business Domain's engine using the datasets created in the data architecture.

- Dashboards are created automatically by the Golden Data Business Domain's engine and using the datasets created in the data architecture, or they are manually created by the data domain owner, depending on the infrastructure definition.

# The Golden Data Platform's Values

Now that there is a general vision of the Golden Data Platform, let's talk about its values as a platform.

### Data Product and Governance-Oriented

The Golden Data Platform is a product that generates and manages data as a product. For it to be true, the data governance must also be embedded in its composition.

This means that the focus must be on the following.

- Effectively managing the Golden Data Platform and the data generated from it by concentrating on features that meet the needs of a large number of people (including customers of the platform and of the data as a product) rather than a few

- Meeting compliance and federated data governance requirements

- Enforcing essential and mandatory requirements to empower and safeguard the data platform and business

- Besides being prescriptive for standards, enable customizations by supporting and encouraging best practices

- It is a data polyglot[2] that supports and manages different possibilities of consumption and persistence from relational, No-SQL, graph, and in-memory technologies that enable batch, streaming, or near real-time as necessary for business needs.

## Source of Truth for Golden Data

Although the Golden Data Platform may generate other types of medallion data, its main goal is to finally generate the golden data.

For this reason, besides leveraging the big data values, the data warehouse ones are also leveraged, totalizing 11 definitions that must be considered.

- **Veracity**: It must ensure the accuracy and authenticity of the data. Ensure useful information, not disinformation.

- **Value**: It must extract and generate value at any time, regardless of its type of data (e.g., historical data, recent data, event-driven, raw or treated, statistical, etc.)

- **Viscosity**: It must push to action due to the criticality that the a-ha information gave.

- **Subject/domain-oriented**: It must be organized and modulated by all domains and its subdomains.

- **Validity**: It must manage data with strong data quality policies, monitoring, and strategies, ensuring all aspects of data quality.

---

[2] Polyglot persistence is the approach of using multiple types of data storage technologies (relational and non-relational) to better attend the demands of data management needs. It was initially raised by Scott Leberknight in 2008, and started becoming famous on the MartinFowler.com blog (2011). Available at https://martinfowler.com/bliki/PolyglotPersistence.html

- **Visualization**: It must ensure efficient monitoring, with quick visualization, and give minimum enough information for further analytics or direct final decision.

- **Integrated**: It must integrate all data sources by consolidating all information and enabling insight into operational processes and new possibilities for leveraging analytics to make decisions and create business value.

- **Time-variant**: It must contain the historical and most current data available for analysis.

- **Velocity**: It must ingest, compute, and analyze massive data faster, considering the necessity of the timeliness of the business requests.

- **Non-volatile**: According to the business needs, it must be able not to change or remove its history from the repository.

- **Volume**: Huge volumes are not mandatory, depending on the business characteristics, but they must be able to store, compute, and analyze big volumes of data and keep their scalability.

## Integrated

Since it is the core of analytics, it must integrate all data sources by consolidating all information and enabling insight into operational processes and new possibilities for leveraging analytics to make decisions and create business value.

The Golden Data Platform must do the following.

- Be business domain/subject modular, giving freedom to the team owners of business domains to implement and continue their functionalities and use or combine capabilities as appropriate for their needs once they are compliant with essential requirements defined for the platform related to data management and data governance.

- Have the capabilities to compute and analyze different types of data: structured, semi-structured, and not structured as necessary for the business contexts (e.g., texts, social relationships, images, videos, etc.)

- Provide explicit and consistent APIs or other mechanisms for integrating and automating features into processes inside and outside the platform.

## Automated

As a platform, it provides automated functionalities for any necessity.

- It defines standards and automate the processes by creating functionalities connected to the data product and governance-oriented, as well as a source of truth for golden data and integrated values.

- If it's not possible to fully automate from the beginning, start with a semi-automation approach and keep the data product management evolution through the minimum viable product (MVP)[3] strategy by continuously improving the product.

- It leverages DataOps to its fullest.

## Easy to Use

It needs to be easy for all customers to use: of the Golden Data Platform, and also for the final consumer of the data generated. As a platform and final data as a product, it provides a:

- User-friendly and intuitive experience of usage of the product that aligns with all the four values: **data product and governance-oriented, source of truth for golden data, integrated** and **automated**.

- Self-service experience to improve business processes, reduce operational effort and frustration, and increase efficiency in cognitive efforts and final results.

---

**Note**   Refer to Chapter 3 to recap definitions related to data consumption and data production.

---

[3] MVP is a strategy for building a simplified version of a product, to test the concept of a product with reduced effort and resources. Once launched the product, evaluate the results of it, learn from customer's feedbacks opportunities of improvement, apply the improvements, and keep the cycle.

With clarity about the Golden Data Platform's values, it's time to talk more about the two main responsibilities of the platform: consume data to produce data as a product and produce the data as a product.

## Data Consumption from Data Services

Data consumption is fundamental for data production; without it, we don't have the raw material to manufacture our product. This raw material can already be golden, or be silver, or bronze data.

Golden and silver data are supposed to already be inside the Golden Data Platform and integrated into all its components. Therefore, the integration with them is easier.

Besides, here, our focus is only on the consumption of the data to produce our golden data, which is part of the downstream pipeline by an organization's holistic vision.

Additionally, we may need to consume bronze data, which can also demand the ingestion process in case they are not on the platform already. Then, pulling data from the data sources (via batch or streaming processes) is required.

---

**Note**    Although the process of consuming data from a source is contemporaneously called the data ingestion process, it's similar to extraction from the traditional environment. However, it uses different technologies to apply similar objectives, usually different ones than was possible to use in the past, as unique solutions were able to integrate the whole data pipeline, and only structured data were the data sources.

---

Here, a huge and very important work must be done together with the domain service owners. It might be the same team or not. It'll depend on the business domain, which can be the same or not, and also on the organization organogram and strategy of the company.

Anyway, it's fundamental to have integration and collaboration among the teams.

Let's start practicing the Golden Data Platform's values and DataOps principles.

## Data Contracts as Key Tools for Data Quality

As you've seen so far, the main root of data quality issues are mainly related to two main reasons: data origin (the data used as a source for the analytics) or related to the data consumed from data produced.

If there's any issue on one of these two fronts, then, we will have a big problem with our data product and will create cascaded problems for the business operation.

For this reason, the Golden Data Platform handles it as well, even though the teams are potentially different. The ingestion process must be implemented in partnership with the team responsible for the operational services implementation, which is the data source, and the business domain's data owner team. Anyway, it's fundamental to have the creation of it for any potential new source. Thus, you may not be the owner of it, but you should be the co-owner of it.

The main concerns related to data consumption are the lack of clear data ownership, technical and non-technical agreements for the consumption of the data, and confidence in the data, which is related to observability.

For this reason, let's talk about what must be defined and followed for data consumption.

### When necessary ingest data

- **Operational services and data ownership**: It's
  critical to clearly define domain ownership inside
  the organization, regardless of the organization's
  strategy. For this, besides these clear definitions

of who is the owner of what, work agreements are fundamental. Agreements that define scopes of responsibilities to ensure the availability and resolution of issues are the foundation for the work (e.g., SLA contracts). In short, it should be considered that any bug or potential issue identified in the operational service or by the data is addressed by the owner team, also following specific contracts.

- **Data observability**: The main concern is confidence that the data consumed is accurate. For this, it's critical to have a clear visibility of the status of the data. For this reason, it's critical to ensure data quality monitoring. The contracts are fundamental. Still, the focus on observability is to visualize the data. The alerts and dashboards must be used in the data observability product of the Golden Data Platform adherent to its module. In other words, by its domain.

  Here, the Shift-left concept is a key approach for data observability and can be performed via streaming or batch computing processes. It must fit the behavior and characteristics of the data.

- **Data contracts**: The agreements are fundamental to address, especially the schema definitions (content, schema, data types, consistency, data freshness), data quality (general data quality and business checks), versioning, security, and privacy restrictions.

## When Just consume the data

- **Data correctly allocated and configured in the data architecture**: Once the data allocation in the data architecture is a key definition inside the platform, tied by the Golden Data Engine, it just needs to be appropriately configured in the platform.

- **Ownership is correctly defined in the Golden Data Platform**: Once the ownership is a key definition inside the platform, tied by the Golden Data Engine, it just needs to be appropriately configured.

- **Available observability for evaluating the sources consumed**: As the observability is already implemented, the product is expected to be available for customers to visualize the data.

- **Data contracts**: Agreements are fundamental to address, especially for security and privacy, as well as the health status of the data for usability, versioning, and SLA.

## Designing Data Contracts

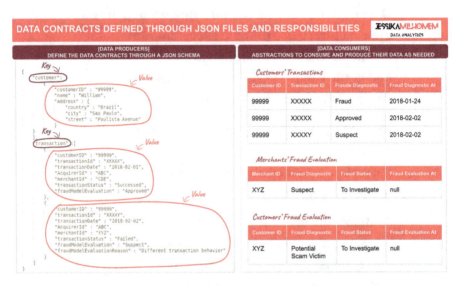

**Figure 5-3.** *Data contracts defined through JSON files, and responsibilities for data producers and data consumers*

The data contracts can be designed using different approaches. To recap it, please see Chapter 3.

Regardless of the approach used, it's important to have the goal clear to ensure the agreements are realized. In a way, it's also possible to apply DataOps best practices.

For this, let's look at a JSON definition of a contract for consumption (see Figure 5-3).

Do the following to produce the data contract.

- Ensure standard definitions for the schema.

- Consider all attributes and specifics necessary for the domain without dependency on detailed definitions from consumption data.

- Enable the consumers to produce their data in whatever way is necessary, with simplified.

Figure 5-3 shows a JSON file with granular information from transactional data that contains all service data produced and available for consumption as consumers need.

Thanks to its schema definition, it's possible to be confident that any new transaction is considered in the contract. Once Golden Data Engine's components are implemented, it'll automatically feed the Golden Data Platform with a transparent approach.

Remember that the Golden Data Platform can leverage the data infrastructure definitions from Lambda or Kappa architectures to feed any data consumer analytics needs for batch and streaming purposes.

## Data Production of Analytics Data Inside the Golden Data Platform

The other fundamental part of the Golden Data Platform work is data production. Data production is related to the T (transformation) and L (loading) of the ETL process. This is when the data as a product is manufactured using the Golden Data Platform.

Just one important disclaimer before moving on. Don't forget that although the focus here is to produce the data, it'll be consumed by others. You may also consume your own data as a product.

So, in data production, you must be double committed.

Besides the data contracts, as you can see in Figure 5-4, the platform mainly comprises four products: Golden Data Engine, data architecture, datasets, and data observability. Each one of them is described in the following sections.

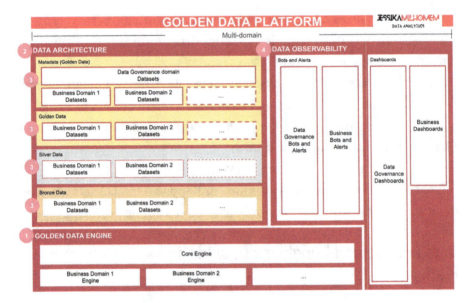

*Figure 5-4.* *The Golden Data Platform: data production*

Although it also contains bronze and silver data, this platform strongly focuses on delivering golden data and full data management for the data inside it because data as a product is also consumed and an important source of golden data.

It has its own values, as shown in the next section. While it keeps the big data's values, it also leverages Inmon's data warehouse definitions to implement a data lakehouse.

Nonetheless, its design follows Kimball's paradigm that defines the data warehouse as a composition of data marts. So, each domain is a piece of this data lakehouse.

It also aligns with the data mesh concepts of domain-oriented ownership, which is related to the subject area definitions of data marts but enriches the ownership details, considering our reality in the big data era.

In short, the Golden Data Platform leverages data warehouse main definitions to design a data lakehouse with features to fully manage the data, ensuring federated definitions of the company and managing requirements necessary for big data circumstances.

The platform's design doesn't focus on the infrastructure environment because it is not a purpose. It is a prerequisite. And we already discussed fundamentals that you should use in case you are also responsible to analyse and define.

Coming back to the Golden Data Platform, there are some approaches that can be considered for its design with data lake components and technologies definition:

## Together with the Data Lake components and technologies definition

You can design your Golden Data Platform when you start defining the architecture in your organization: the repositories you'll work with, how to compute your data, what services need to be implemented, and so forth.

This would be the ideal moment for it because then, it'd be possible to align details for the federated definitions along the company in a fashion that some customized infrastructure definitions are implemented in the company-level/federated infrastructure definitions by the enterprise central team (if you have this team organization in your company).

We'll talk more about it in the next sections.

## Upon the Data Lake Components and Technologies Definitions

The other possibility is to implement the platform upon one existing company-level infrastructure with already federated functionalities for a company.

If I bet which option is the more common, it would be this one; the opposite would be for new startups.

It's time to implement the platform upon which an environment is already set up, with repositories defined, services hired or developed, and contracts signed with suppliers.

So, in terms of the design, it's the scenario to perform the Lego exercise. In other words, it's time to analyze the necessities for the platform and how to leverage the solutions designed and integrate them with the platform.

In terms of amplitude scope, the Golden Data Platform can be designed at two levels.

- **Organization level**: In this approach, the platform is designed, implemented, and managed along the organization, by a centralized technical organization.

- **Macro-domain level**: In this approach, the platform is designed, implemented, and managed inside a macro-business domain area (usually business units, such as marketing, fraud, finance, etc.)

## Four Products that consists the Golden Data Platform for Data Production

You have now had a comprehensive overview of the Golden Data Platform and its main pillars that must be considered for each one of the products that makes it up. Thus, let's discuss each one of them!

## The Golden Data Engine and Its Modules

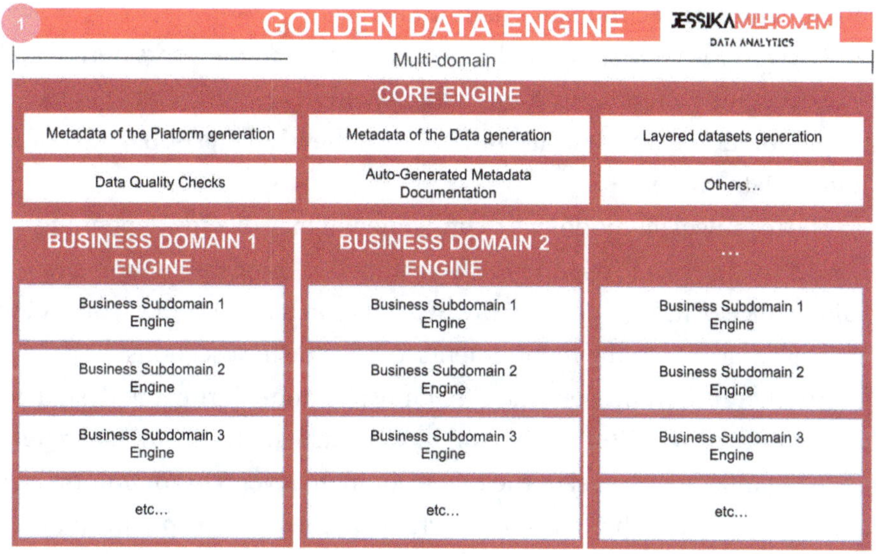

*Figure 5-5.* *The Golden Data Engine*

The Golden Data Engine is a product that contains the main business intelligence inside it. It is responsible for automatically orchestrating and generating the datasets in the data architecture based on standard definitions for the platform.

The engine is the heart of the platform. The automation lives mainly inside it. The Golden Data Engine comprises the core and business domain engines (see Figure 5-5).

Regardless of when you start to design the Golden Data Platform (together with the data lake components and technologies definition, or upon the data lake components and technologies definitions) and the technologies decided by you or "decided for you," the approach is still the same: it'll be designed a framework, and then, the services of it which composes the operationalization of the engine.

241

The framework can be designed and implemented during the data lake components and technologies definitions or upon the data lake components and technologies definitions.

Anyway, the Golden Data Platform comprises frameworks performed by services (as components[4] or as an application system[5]) that are one unique solution to orchestrate and perform the whole responsibility of the Golden Data Engine.

The technical methodology to implement it may diverge depending on when you design it, the approaches defined for implementation and maintenance of it, and the structure of teams you have in your organization. We'll talk more about the requirements for it in the next sections.

Another important definition that must be done during the design is regarding the amplitude scope of the Golden Data Platform. It can be at the organizational level or the macro-domain level. This definition is important to establish not just the technical aspects of the platform but also the non-technical ones. We'll talk more about the requirements for it in the next sections.

## Core Engine

The core engine is responsible for applying domain-agnostic intelligence with standard definitions and features applicable to a huge number of domains if not all.

It's composed of different components related to the functionality of the data platform itself and the application of cross-definitions for the domains, such as functionalities for data governance aspects such as data quality controls.

---

[4] A software functionality or a set of functionalities that accomplish a particular task or purpose. It has no user interface and can be executed in parallel to other services.

[5] A group of hardware and software items created to satisfy one specific business application necessity.

All features must follow the defined scope amplitude at the organization level or in a macro-domain area to ensure the reliability[6] of the platform and avoid duplicated effort and services inside the platform (see Figure 5-5).

The goal is to ensure that everything that is cross-cutting and necessary for the business (at an organizational level or macro-domain perspective) is here. It performs responsibilities or has functionalities leveraged or inherited by other business domain engines.

Some examples of functionalities that need to be implemented in the core engine.

- **Metadata of the platform generation**: It's responsible for mapping and storing data and logs for the generation of datasets through the platform, such as which domain engine is being used, when, by who, and errors generated by the engine itself. Considering the approach of implementing frameworks, here we have one abstraction responsible for that.

- **Metadata of the data generation**: It's responsible for defining functionalities to map and store data about the data being generated, such as the team owner of the data, its direct sources, the business context of it, and so on.

  Considering the approach of implementing frameworks, here we have one abstraction responsible for that.

- **Layered datasets generation**: It's responsible for defining all the generic, domain-agnostic rules that are necessary to implement the data in the appropriate layers of the data architecture, regardless of its domain.

---

[6] Practices to avoid failures of a service, system, or product.

Considering the approach of implementing frameworks, here we're going to have one abstraction responsible for that, which is used by the business domain engine that inherits the class/object to perform the specifics of the domain.

- **Data quality checks:** It defines all the generic, domain-agnostic rules necessary for data quality checks. Validate if the data is duplicated for the final datasets composed, which cannot be approved for golden data, or implement integration with any eventual tool that the organization uses for and enable usability by each business domain, without the necessity of duplicated integrations, and so forth.

  Considering the approach of implementing frameworks, here we will have one abstraction responsible for integrating with any third-party solution as part of the infrastructure or in-house solution. If it doesn't exist, we'll have features implemented inside the Golden Data Platform inside the core engine. Once implemented, it is used by the business domain engine that inherits the class/object to perform the specifics of the domain.

The core engine should have functionalities that are specific for silver data, besides the golden data, as it's expected to generate them for intermediate datasets generated or for users' usage as A/B test scenarios.

Besides, although not focused on that, the engine might also have functionalities to generate and manage bronze data. It'll depend on the scope amplitude defined during the design of the solution.

In terms of ownership, the core engine's functionalities should ideally be managed by a team with horizontal responsibilities at an organizational level or a macro-domain level.

This team should be responsible for defining and applying the standard and essential requirements for the engine, sharing the best practices through some mechanism, and supporting the users in using it independently.

As the owner of the core functionalities, the central team is also responsible for ensuring full integration and continuous communication with the customers.

## Business Domain Engine

The domain engine is business-driven and where the business intelligence lives. It's responsible for creating the data within the data architecture, with standard schemas and data intelligence definitions for the domain in an automated fashion.

While also enabling flexibility for customizations by specific contexts necessary for the business (see Figure 5-5).

It's composed of functionalities for all subdomains of that specific business domain. Or it can have functionalities for the domain as a whole and perform specific responsibilities for it.

All features must follow the Golden Data Platform values.

For integration with company-level or macro-domain technologies, it's fundamental to integrate with the core engine and always, as necessary, inherit the functionalities and use them as needed for the business domain.

Business domains that are cross-cutting and necessary for the business (at an enterprise level or macro-domain perspective) are here and must be owned by appropriate teams that perform similar cross-cutting responsibilities. Otherwise, it must be owned by business domain teams. Defining ownership must be part of the metadata of the dataset and other characteristics.

Some examples of business domains and their subdomains that could be implemented in the business domain engine.

- **Fraud domain**: It is a domain that should be owned by the fraud domain team and must contain all the subdomains related to it. Each subdomain might be owned by specific teams inside it or the central team of the fraud business domain.

  - **Fraudsters and suspects subdomain**: The engine must contain the intelligence to identify fraudsters and suspects. It might be done with heuristics, advanced analytics, and/or different technologies. Each organization has its own strategies and approaches, which must be considered in the domain engine to automatically generate the datasets (regardless of the technology used) inside the architecture.

  - **Modus operandi subdomains**: The engine must contain intelligence to identify the modus operandis and apply the rules necessary to automatically generate the datasets in the data architecture (regardless of the technology used for it).

  - Etc.

- **Data governance domain**: It is a domain that should be owned by a central team, according to the scope amplitude definition (at an organization level or macro-domain perspective). The data governance domain must contain all the subdomains related to it and the rationale for them, with a standardized definition regarding its geo-location while also keeping the customization possibilities.

- Data quality subdomain

- Business processes subdomain

- Security management subdomain

- Data cost management subdomain

- Etc.

# Data Architecture and Its Functionalities

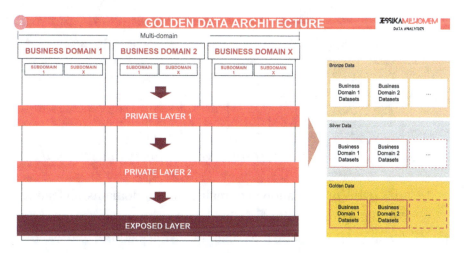

***Figure 5-6.***  *The data architecture*

The architecture is layered in business domains and subdomains and is controlled by data usability and security, also leveraging the  data mesh concept, It is data-medallion oriented, as we discussed at the beginning of the book, and works by splitting the data into three layers: bronze (raw), silver (pre-treated), and golden data (source of truth for business decisions).

All of them are products. However, only the golden data is available for consumption (see Figure 5-6).

The domain engines automatically generate the datasets with the already standardized security controls in the architecture.

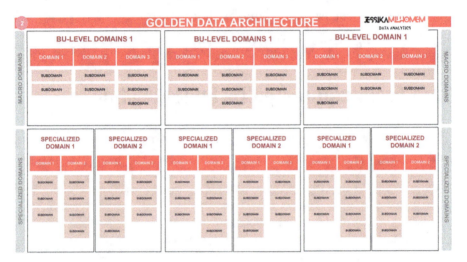

*Figure 5-7.* *The data architecture: organization of data domains*

In terms of the organization of domains and subdomains, all features must follow the scope amplitude defined: organization level or for a macro-domain area to ensure the reliability[7] of the platform and avoid duplicated effort and services inside the platform.

The goal is to ensure that the cross-cutting necessities for the business (at an organization level or macro-domain perspective) are here and can be used by the company regardless of its strategy (see Figure 5-7).

Using the previous examples shared in the business domain engine section, we would have two business unit–level domains: fraudsters and data governance.

---

[7] Practices to avoid failures of a service, system, or product.

The fraudsters domain has at least two subdomains: fraudsters and suspects. There is a modus operandi for domains and specialized domains.

The data governance domain has at least four subdomains: data quality, business processes, security management, and data cost management.

The layers are defined by the usability of the data, not by its granularity, nor specifically by its medal. Moreover, the datasets are classified with a medallion layer, which is part of the usability of the data but not the reason for its definition.

Yet, using the fraud domain as an example, let's define the following layers of data usability.

- **Operational private data**: This layer maintains the data necessary for fraud analytics operations, and only fraud can access that data. For example, a granular list of fraudsters and suspects is very sensible data and must also be used in alignment with data privacy laws. Due to its characteristics, this layer would contain just golden data, which must be accurate.

- **Business exposed data**: This layer maintains just the anonymized data and has the information that other teams need, such as the total of frauds and total amount of losses that happened for a specific modus operandi in a specific period. Due to its characteristics, this layer only contains golden data and must be accurate.

# Data

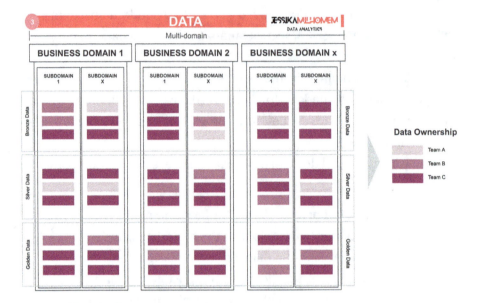

*Figure 5-8.* *The data*

The data is a product by itself, as it contains a proposal of existence to solve a real problem for some customer (regardless the persona of it), and must be designed, implemented, and maintained. Review Chapter 1 to recap data as a product.

All datasets (regardless of format) must follow the golden data values and be generated via a business domain engine to ensure the platform's reliability and avoid duplicated effort. As soon as it's generated, it'll have clear ownership defined according to its business domain or subdomain (see Figure 5-8).

# Data Domains and Subdomains

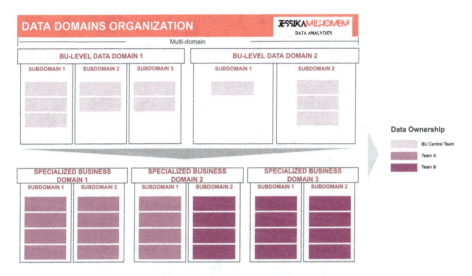

***Figure 5-9.*** *The data domains' organization and ownership*

The datasets must be created and organized by domain. The domains should also be organized by their category inside the domain.

- Business unit domain (or macro-level domain)

- Specialized domain

All the cross-cutting subjects are related to the business unit–level domain and can have related subdomains. They have information regardless of the specialization of the domain. They are actually baseline data sources for the specialized domains.

The specialized domains are responsible for specific business contexts and processes of a business unit domain and have their own detailed subjects.

Macro-level data domains should be owned by a central team inside the business unit that serves specialized teams, considering the defined scope amplitude.

The specialized data domains should be owned by the responsible teams (see Figure 5-9).

Yet, using the fraud domain as an example, we could have different teams in the fraud organization who specialize in modus operandi definitions. A central team is responsible for cross-cutting subjects, such as mapping the subdomains of fraudsters and suspects.

The data generated for the fraudsters and suspects is important information that specialized domains can feed.

# Data Observability

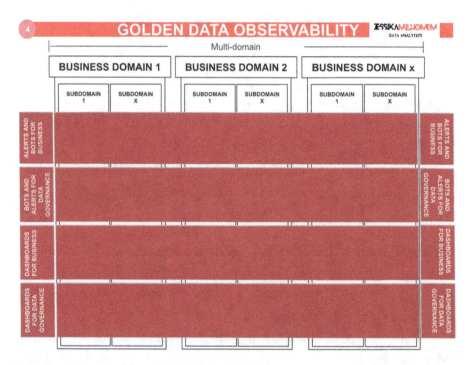

*Figure 5-10.  Golden Data Observability*

Data observability contains all the business and data governance alerts and dashboards for all domains and, if applicable, by subdomains. It's business-driven, standardized, and modular by business domain, like all the other components of the platform. It must follow the golden data values to ensure the reliability[8] of the platform and avoid duplicated effort.

The alert intelligence must be generated by the Golden Data Engine, while the dashboards might be (depending on the technology used by you/your organization). The dashboards and alerts must use datasets generated in the platform, ideally just by golden data. However, depending on the subject context, silver data can be used once medallion identification is generated via a business domain engine (see Figure 5-10).

The BU's central team should own the Golden Data Observability dashboard, considering the scope amplitude defined. However, the business domain teams must have the flexibility and autonomy to create specific and necessary observability.

# Reasons to Use the Golden Data Platform

Let's clarify and make explicit the benefits of using the Golden Data Platform versus not using it.

- Provides a clear and unified architecture that is medaled, governed, layered, and modulated

- Provides a clear, unified, and integrated source of truth for golden data in the organization for any type of customer

- Although unified and integrated data architecture is not centralized and is managed by appropriate data domain owners, adherent to its module

---

[8] Practices to avoid failures of a service, system, or product.

- Easier and faster data governance implementation and maintenance

- Easier and more integrated data governance monitoring

- Healthy and robust scalability

- Modular by domains inside the four products that consists the Golden Data Platform

- All the components attend and fit very well to batch, streaming, or hybrid data needs

- The Golden Data Engine handles the Shift-left approach for batch or streaming

- Capability to automated scale the domains for different geos due to standardizations

- More cost efficiency is due to standardizations and integrated architecture.

The following are possible consequences of not using the Golden Data Platform.

- Inflexible architecture: siloed or centralized

- Inflexible data management approach: siloed or centralized

- Complexity to manage centralized needs

- Lack of confidence in the accuracy (medal) of the data for data consumers

- Not integrated solutions

- Hard and complex data governance implementation and maintenance for siloed data architectures

- Long period and effort to implement and maintain data governance for centralized data architectures

- Complexity for data governance monitoring for siloed data architectures

- Hard effort and a long period of time to implement domains for other geos

- More expensive architecture due to less efficient architecture

# 5.2.  Golden Data Platform Requirements

To design, implement, and maintain the Golden Data Platform, you have to follow and address technical and non-technical requirements.

## Clearly Mapped Business Issues

The problem is the core of everything. It needs to be evaluated and understood, as well as potential customers.

The business issues must be clearly mapped for the project implementation and continuously evaluated during the product's life cycle. It'll be covered in the next sections.

## Sponsorship

As soon as the scope of the potential project to implement the product is clear, it's important to map its sponsor.

Without sponsorship, the possibility of working on it doesn't exist.

## Scope Amplitude

Defining the amplitude of the scope of the platform is fundamental not just for its design and implementation but also for the maintenance and ongoing work agreements with other teams and for ensuring the accuracy, efficiency, and reliability of the data at an organizational level.

The options are organization level and macro-domain level (see Table 5-1).

*Table 5-1.* Scope Amplitude

| | ORGANIZATION LEVEL | MACRO-DOMAIN LEVEL |
|---|---|---|
| **DEFINITION** | The platform is designed, implemented, and managed along the organization by a horizontal central team. | The platform is designed, implemented, and managed inside a macro-business domain area (usually business units, such as marketing, fraud, finance, etc.) |
| **PROS** | • One unique platform that contains all functionalities in one unique core engine<br>• One unique platform that contains all the macro domains in a consistent architecture<br>• Unique team with professionals focused on horizontal needs<br>• Domain-level teams focused on what is strict to their domains but keeping integration and compliance with the organization-level solution<br>• Consistent with the guidelines of subject-oriented from the data warehouse and Golden Data Platform<br>• Consistent with the domain-driven ownership context of data mesh, due focus on company-level or macro-business-level domains. | • More freedom with a controlled backlog for one specific macro-business domain<br>• One unique team to manage macro-domain level needs by ensuring compliance with company-level requirements<br>• Internal central team to manage across business-level domains<br>• Domain-level teams focused on what is strict to their domains but keeping integration and compliance with the organization-level solution<br>• More time available for innovation at the company level<br>• Faster in resolving issues and failures<br>• Consistent with the guidelines of subject-oriented from the data warehouse and Golden Data Platform<br>• Consistent and fully fits with the domain-driven ownership context of data mesh |

*(continued)*

*Table 5-1.* (*continued*)

| | ORGANIZATION LEVEL | MACRO-DOMAIN LEVEL |
|---|---|---|
| **CONS** | • Large volume of horizontal requests for one backlog<br><br>• Dependency of unique team company-level to manage macro-business-level functionalities<br><br>• Dependency of one unique team to manage company-level domains<br><br>• Reduced time available for innovation at the company level<br><br>• Potential delay in resolving issues and failures | • Eventual duplicated functionalities for the core engine, when evaluated at enterprise level that is unique when evaluated at a macro-domain level<br><br>• More core teams inside macro-business domains, additionally to an eventual central team to manage cross-needs |

What's the ideal approach if both have pros and cons?

It depends. And of course, you need to evaluate your context, but here are some points to consider.

If you're a small/mid organization, you may have the following characteristics.

- Small team

- Small volume of data or small volume of business contexts

- Limited resources

If so, the organization-level approach is probably the best approach for you.

If you're a growing organization, you may face the following characteristics.

- Bigger team

- More flexibility for different resources

- A large volume of data and business contexts

If so, the domain level is probably the best approach for you.

You may also have different characteristics, even considering your size. For instance, you are small but in an exponential growth moment, or you're big but in a recess time. Then, the merge of them, accordingly to work-agreements and ownership definitions may be the best for you. It's important to evaluate all the characteristics and consider the best approach. Let's go deeper on it.

# Scope Definition and Work Agreements Among Personas

Regardless of the approach you use during the project (waterfall, agile, or mixed) and how you organize the product management of the data products, it's critical to define clear scope and work agreements among the personas.

Table 5-2 describes the scope and work agreements to be considered for the project moment and the product life cycle.

*Table 5-2.* *Scope Definition and Work Agreements Among Personas*

| | SCOPE | AGREEMENTS |
|---|---|---|
| **SPONSOR** | High-level manager responsible for ensuring the project is properly funded to solve relevant business problems or needs and support the data product manager and team to put it in place | **Before and during the project**<br>• Ensure the project is properly funded<br>• Lead the project by communicating with the high-level management about the project to ensure engagement and formal authorization<br>• Provide a significant role in the development of the initial scope and charter<br>• Serve as an escalation path<br>• Authorize changes in scope<br>• Provide go/no-go decisions<br>**During the product life cycle**<br>• Lead the project by communicating with the high-level management about the project to ensure engagement and formal authorization<br>• Serve as an escalation path<br>• Authorize changes in scope<br>• Provide go/no-go decisions |

*(continued)*

*Table 5-2.* (*continued*)

| | SCOPE | AGREEMENTS |
|---|---|---|
| **DATA PRODUCT MANAGER** (a.k.a. as project leader during the project) | Senior data manager responsible for solving relevant business problems and needs through short, mid, and long-term product vision for data | **Before and during the project**<br>• Map the problems and business opportunities to be addressed<br>• Design the project scope and short, mid, and long-term vision and align with the sponsor for validation<br>• Plan, execute, and conclude the projects according to the scope, budget and deadlines<br>• Manage the project team and allocated resources<br>• Monitor project progress and identify potential risks<br>• Report project progress to the sponsor and stakeholders<br>• Support to the team to refine the technical needs<br>**During the product life cycle**<br>• Strategic focus<br>• Product vision<br>• Competitive analysis<br>• Product portfolio management<br>• Continuous monitoring of products<br>Also part of the role scope, ideally performed by less seniors Data Managers: |

(*continued*)

*Table 5-2.* (*continued*)

| SCOPE | AGREEMENTS |
|---|---|
| | • Prioritization of initiatives |
| | • Refinement of the business and technical needs |
| | • Collection of feedbacks |
| | • Manage the professionals |
| | • Support, guide, and collaborate with less senior data managers to plan and execute the project activities |
| | Also part of the role scope, ideally performed by less senior data managers |
| | • Tactful focus |
| | • Backlog management according to the senior data manager (a.k.a. data product manager) |
| | • Deliverables |
| | • Close relationship with the dev team |
| | • Support to the team to refine the technical needs |
| | • Write the (user) stories |
| | • Manage the professionals, the agendas, and the detailed communication among the team |
| | • Monitor project progress and identify potential risks |
| | • Constantly communicate with the more senior data manager (a.k.a. data product manager) |

(*continued*)

261

*Table 5-2.* (*continued*)

| | SCOPE | AGREEMENTS |
|---|---|---|
| **STAKEHOLDER/ CUSTOMER** | Who directly suffers from the business problem, or who has the direct business needs<br><br>It can be an internal (professional with business scope inside the organization) or an organization's direct customer | **According to the type of customer** (internal or company level)<br>• Define and share the business goals and needs to compose the scope of the project<br>• Share feedback and do evaluations and validations (a.k.a. UAT - User Acceptance Testing)<br>• Actively participate in the allocation of resources<br>• Share potential risks for the business and the project during the whole journey, but especially at the beginning<br>• Advocate by the project |
| **TECHNICAL TEAM** | The professionals responsible for operationalizing the solution by coding the data product | **During the project and product life cycle**<br>• Execute the hands-on work for the project<br>• Report the progress of the tasks<br>• Create a plan for the sprint: the sprint backlog<br>• Instill quality by adhering to a definition of done<br>• Adapt their plan every day toward the sprint goal<br>• Hold each other accountable as professionals |

# Work Agreement on Data Platform Ownership

Although it might differ in some product life cycle details, the general perspective is kept for the ownership of the data platform components (see Table 5-3).

It's important that you do it before the project to set expectations, after its finalization, and eventually, during the product life cycle of the project.

***Table 5-3.*** *Work Agreements on Data Platform Ownership*

| | SCOPE | TEAM PROFILE | AGREEMENTS |
|---|---|---|---|
| **CORE ENGINE** | Ensure platform and cross-level needs are embedded as functionalities to perform specific responsibilities by themselves or to enable the domain engines to use them for their own needs | Central team (organization level or macro-domain level) | • Compliance with Golden Data Platform values<br>• Compliance with federated data governance policies<br>• Compliance with the essential features of the core engine<br>• Map, implement, and ongoing the common functionalities necessary for the scope amplitude defined |

(*continued*)

*Table 5-3.* (*continued*)

|  | SCOPE | TEAM PROFILE | AGREEMENTS |
|---|---|---|---|
| **DOMAIN ENGINE** | Ensure domain-level needs, standards, and intelligence are embedded as functionalities to perform specific responsibilities by themselves while also enabling customizations | Domain teams and (if applicable) Central team (organization level or macro-domain level) for cross-domain | • Compliance with Golden Data Platform values<br>• Compliance with federated data governance policies<br>• Compliance and integration with the core engine features that contain the essential requirements for the Golden Data Platform<br>• Map, implement, and ongoing the common functionalities necessary for the domain specified<br>• Implement ongoing customizations |

(*continued*)

*Table 5-3.* (*continued*)

|  | SCOPE | TEAM PROFILE | AGREEMENTS |
|---|---|---|---|
| **DATA ARCHITECTURE** | Ensure the architecture is business domains and subdomains layered, as well as layered by data usability and security control | Central team (organization level or macro-domain level) | • Compliance with Golden Data Platform values<br>• Compliance with federated data governance policies<br>• Attend the essential requirements of the core engine<br>• Compliance with the minimum features of the data architecture<br>• Map and implement ongoing common functionalities necessary for the data architecture |

(*continued*)

***Table 5-3.*** (*continued*)

|  | SCOPE | TEAM PROFILE | AGREEMENTS |
|---|---|---|---|
| **DATA** | Ensure the shared data is healthy in the business and technical aspects are continuously accurate, reliable, and automated in the platform components | Domain teams and (if applicable) Central team (organization level or macro-domain level) for cross-domain | • Compliance with Golden Data Platform values<br>• Compliance with federated data governance policies<br>• Have the intelligence generated through the domain engine<br>• Ensure the shared data is healthy, updated, and automated within the domain engine<br>• Ensure the shared data is healthy in technical aspects by ensuring the performance, reliability, and data quality daily-basis follow-up through data observability |

(continued)

*Table 5-3.* (*continued*)

|  | SCOPE | TEAM PROFILE | AGREEMENTS |
|---|---|---|---|
| **DATA OBSERVABILITY** | Ensure the alerts and dashboards are healthy in the business and technical aspects by keeping them continuously accurate, reliable, available, and automated in the platform components | Central team (organization level or macro-domain level) for cross-domain and Domain teams for customized data observability | • Compliance with Golden Data Platform values<br>• Compliance with federated Data Governance policies<br>• Map, implement, and ongoing the common functionalities necessary for the data architecture |

# Infrastructure as a Service

The Golden Data Platform is a service platform focused. Thus, the infrastructure must be offered as a service for use, not necessarily related to a cloud service but as a service for the platform. This means that a specific team responsible for it should offer it at the company level. As the goal of the Golden Data Platform is to manage data inside a data lakehouse, it doesn't make sense to have silos of infrastructure for it.

Nonetheless, depending on the design, it can diverge a bit.

- **Designed during the definition of data lake components and technologies**: It can be implemented using any of the scope amplitude definitions.

    - If defined to have all the company-level capabilities, the team responsible for it can also accumulate other responsibilities that are related to infrastructure, such as the definition and maintenance of infrastructure services to ensure the environment performs the core engine responsibilities.

- Similar responsibilities can sit on the macro-domain teams if it's defined as having a macro-domain approach. However, I wouldn't recommend it, as the Golden Data Platform works with a data lakehouse, and for this reason, it should ideally have similar infrastructure for it, due to similar business needs, and keep infrastructure specifics just for business specifics depending on the domains.

- **Designed upon the data lake components and technologies definitions**: For this scenario, the main definitions for the data lake components, such as infrastructure services, were already defined. Besides, it can be implemented using any of the scope amplitude definitions. Thus,

  - If defined to have all the company-level capabilities, it'll be important to revise and align responsibilities related to the Golden Data Platform, which were previously addressed in another way, to be embedded in the core engine as features and integrated as responsibility for the team.

  - If it's defined as having a macro-domain approach similar to the organization level, it must be revised and aligned to fit the macro-domain-level context.

Regardless of its definition, it is important to have one central team responsible for cross-cutting definitions and infrastructure needs, regardless of the use of cloud services.

It's fundamental to have professionals focused on the data infrastructure that your company has chosen for its infrastructure at the company level.

In short, the data platform admins' responsibilities must be kept.

If there is any necessity for a specific business context related to it, then the data platform admin for the new technology specific to this need should be responsible for it inside the macro-domain level. If the new business requirement becomes popular, consider moving it to a central team to operate it from an organization-level approach.

## Federated Data Governance

Having one central team that works according to the federal definition of data governance within the organization is fundamental.

The teams working on the Golden Data Platform (regardless of the scope amplitude) must ensure compliance with them.

## Promotion of a Data Definition

Besides, the federated definitions of data governance should be considered for any eventual policy to define the medal for the data (see Chapter 3). It's fundamental that inside each macro-domain, the data is clearly defined and complies with the business definitions for the domain.

The data product manager of the Golden Data Platform must design this definition.

# 5.3. Data Product Management in a Nutshell

This section discusses the macro steps to design and implement the Golden Data Platform. Actually, the steps are the same regardless of the data product type you will work with: data as a product or data product. I'll discuss each one of the steps in the upcoming chapters.

First, it's important to remember that the best dynamic for an efficient product design, implementation, and rollout is to start with a project. Then, as soon as the product is ready to be released, move it to the ongoing, with business-as-usual responsibilities. And then keep the cycle.

## Designing the Data as a Product: Learning Phase

The first step of the process is to learn everything related to the business context and the issues related to that. It's the learning moment!

Of course, the full journey enables you to learn, and you should be focused on that.

In this phase, you learn about the business processes, the whys, and the problems.

Once you have more clarity about the business context, you'll define the success criteria for the project.

It's fundamental to have clear expectations for it. Of course, to define it, you'll need to learn more details about a process and use it as a use case to solve problems.

Finally, define the deliverables during the project.

Each step is fundamental and sequential to enable the design of a good strategy, which is validated with a tactical plan and execution in a more agile process.

## Designing the Data as a Product: Exploration Phase

With clarity about the problems the business faces, studying and defining the project's success criteria, and the expected deliverables, it's time for the exploration phase.

In this exploration phase, your goal is to design the ideal solution by prototyping it based on the definitions realized there, and ensuring all the definitions and requirements we alined thus far.

With the prototype realized, test the results and validate if they relate to the design defined.

Adjust and redo the same macro step if it isn't as expected. Eventually, even coming back to the previous step.

# Designing the Data as a Product: Materializing Phase

As soon as the whole exploration is finalized successfully, implement the solution and put it into production.

With the product in production, it's fundamental to map and measure the results of it.

Again, it's fundamental to iterate and validate if the results are positive and as expected and to validate if there's any opportunity for improvement.

The project itself is finalized, and if there's any opportunity for improvement, it moves to the backlog, and the cycle continues.

However, before moving on to the next improvements, it's fundamental to ensure the maintenance of the product.

# Ownership Process/Continuous Cycle/ Ongoing

The maintenance of the product must ideally be part of its development cycle.

However, the reality is not so simple. Thus, it's fundamental that you prepare yourself and your team to support your customers.

Otherwise, imagine the frustration you're going to create with them when something fails and no one is taking care of it because you're exploring other new opportunities!

That's right! Don't be that kind of product manager. Be a good one!

Thus, ideally, this phase, which is the last macro phase, must be planned and designed before the product is released (please!).

Define the product's maintenance process and the next steps. Include the continuation of the cycle by ensuring the product is completely ongoing.

Of course, mapping and measuring the maintenance and ongoing maintenance are fundamental for this work.

# 5.4. Summary

I'm super excited that we finally got here! This is a special chapter for me, and I hope it is for you also!

Let's recap the main takeaways of this chapter.

- The **main business problems** that incentivized the creation of the Golden Data Platform are necessary to address the following.

  - Low business ownership and expertise for cross-subjects

  - Lack of modularity and governance

  - Lack of standardizations

  - Low efficiency for data factoring and management

  - Lack of ownership visibility

  - Lack of efficient observability

  - Complex and non-integrated maintenance processes

  - Lack of data governance integration with the business analytics operations

  - Inefficient data cost management

- The **Golden Data Platform** is a data product that
  generates and manages data as a product. It's business-
  driven, designed, and layered by domain, usability, and
  access control to deliver data as a product. The Golden
  Data Platform delivers data as a product with modular
  intelligence, solving and simplifying data management
  to create and maintain the ecosystem.

- It is made up of four main products.

  - Golden Data Engine

  - Data architecture business-driven, domain,
    usability, and security layered

  - Data

  - Golden Data Observability

- **The requirements** for the Golden Data Platform are as
  follows.

  - Clearly mapped business issues

  - Sponsorship

  - Scope amplitude

  - Scope definition and work agreements among the
    personas

  - Work agreement about the data platform
    ownership

  - Infrastructure as a service

  - Federated data governance

  - Promotion of the data definition

- **Data as a product** starts with a project, then, once finalized with success, it includes another approach to ensure the complete ongoing of the product.

## Summary

In this chapter we talked about the Golden Data Platform, what it is, which problems it solves, and how to implement it. Next, let's talk about the practical approach to managing data products regardless of the type of it: data product or data as a product.

The Golden Data Platform is also included here because it relates to both product types.

# CHAPTER 6

# Data Product Management

Let's now focus on how to put data product management into practice. The remaining chapters talk about enabling the management of any data product regardless of its type. Of course it also includes the Golden Data Platform we have seen in the previous chapter.

This chapter discusses the data product management framework and the canvas I created to guide product data management.

As I mentioned, there are some frameworks in marketing for product management, and new ones are always emerging. This book doesn't use any of the existing ones. I'm going to use the one I designed by following its quadrants. I call it **Data Product Management Canvas**.

This is because although there is a good canvas in the market, I've always had difficulty fitting it for data product management.

Besides, for technology in general, faster answers are important, but when we talk about data, it is even more important.

This framework works for other product contexts, not just data products. But it's not the focus of this book.

Before moving on, I want to note that this chapter is the foundation for the remaining chapters.

© Jessika Milhomem 2025
J. Milhomem, *Data Product Management in the AI Age*,
https://doi.org/10.1007/979-8-8688-1315-3_6

# 6.1. The Data Product Management Canvas

A canvas is a visual tool for analyzing and organizing ideas for innovation by creating business value.

With that in mind, I created the Data Product Management Canvas that consolidates the framework I designed.

The Data Product Management Canvas comprises two main pillars: Business Driven and Product Driven (see Figure 6-1).

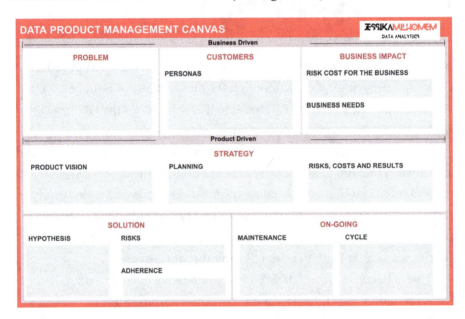

***Figure 6-1.*** *The Data Product Management Canvas*

The Data Product Management Canvas represents the process of managing data products, as discussed in the upcoming chapters.

The canvas contains twelve quadrants to be followed: four are in the Business Driven pillar, and eight are in the Product Driven pillar.

# The Business Driven Pillar

The Business Driven pillar is focused on understanding the problem or challenge faced, the business impacts, and mapping the potential customers.

The approach consists of mapping the problems or challenges, considering its own characteristics and impacts on the business, and mapping the individuals involved, the potential customers.

This is the fundamental pillar for the whole work to be done forward. Without this pillar, it's impossible to design and implement any product.

It can be done, but the developed product may not solve any problem and cause consumers to consume it. However, that's not the idea, right? Otherwise, we'll create potential losses instead of benefits for the business.

# The Product Driven Pillar

The Product Driven pillar is where the whole product concept is designed, implemented, released, and managed.

It is completely dependent on the Business Driven pillar because it is where the business problem and context are mapped. These are fundamental materials that enable the design and factoring of the product.

The approach consists of designing and implementing the strategy for the product, the solution, and the ongoing.

The strategy category considers the product vision, the strategy planning, and the related risks and costs.

The solution category considers the hypotheses, the risks related to its resolution, and then the adherence to the product.

The ongoing category considers the maintenance and cycle rationale.

The Product Driven pillar consists of strategic and tactical approaches. Consequently, it guides the operational actions.

# 6.2.  Summary

Let's recap the main takeaways of this chapter.

- A canvas is a visual tool for analyzing and organizing ideas for innovation by creating business value. And it is an excellent tool for product management if it follows a framework.

- A Data Product Management Canvas is structured to enable the execution of the data management product based on the framework I created, which is described in this book. It has two main pillars: Business Driven and Product Driven.

## Summary

In this chapter we discussed the framework and Data Product Management Canvas, including what it is and how to use it.

Next, let's talk about each part of this canvas and how to put it into practice to manage data products regardless of the type—data product or data as a product—including the Golden Data Platform.

# CHAPTER 7

# Designing the Data Product: Understanding Phase

This chapter discusses the understanding phase. For this, we'll cover the Business Driven pillar's quadrants from Data Product Management Canvas I created (see Figure 7-1).

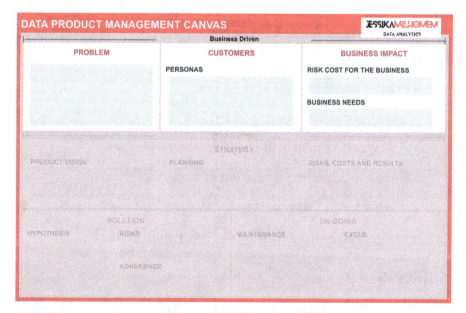

***Figure 7-1.*** *The Data Product Management Canvas: Business Driven pillar*

© Jessika Milhomem 2025
J. Milhomem, *Data Product Management in the AI Age*,
https://doi.org/10.1007/979-8-8688-1315-3_7

I'll do it pragmatically by covering a dynamic process to understand the problems and mapping the risk costs for the business, the potential customers, and expectations regarding resolutions.

After reading and learning everything described in this chapter, you can use the Data Product Management Canvas to plan and visualize it.

# 7.1.  Understanding the Problems

## Mapping a Problem

The beginning of anything starts with the problem definition. Thus, the first step is to map the problem or challenge for the business.

The Business Driven pillar is focused, as the name says in the main business context.

Of course, as a framework, it can include as many questions for definitions as necessary for the business context in which you're going to work.

However, I'd summarize the most important questions to be answered as follows.

- What is the problem or challenge?

- Who is mainly impacted by it?

- How is it a problem or challenge for the business?

- How much does it cost to not do anything?

- What are the expected results if we work on resolving it?

These questions touch all the blocks (see Figure 7-1).

## Problems

All the preceding questions must be addressed to map the scope of the discussion.

However, the best is to split the problem into pieces.

In the Problems quadrant, the main goal is to understand the problem in the business. It can also be a challenge faced.

I'm talking about any problem: direct for the business or technical, but impact the business. It can be related to internal or external issues derived from the market context, such as a competitor or a regulation/legal context.

It can be any size by itself and also for the big picture. If you see it's a big problem, the idea is to keep the same approach mentioned earlier: split it into pieces and understand its complete context and trajectory.

The process to learn about the problems is by performing interviews with the key users who know the characteristics of the problem you're discovering. The goal of this process is to conclude with a clear definition of the faced problems that need to be solved. See Figure 7-2.

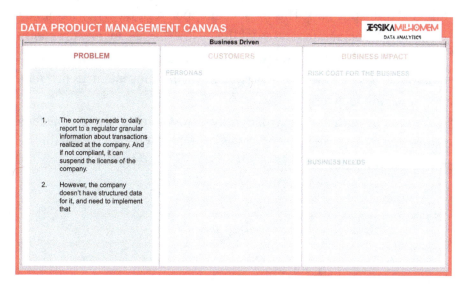

***Figure 7-2.*** *Data Product Management Canvas: Problems Sample*

# Customers: Personas

For the Customers quadrant, there are two moments for the discovery.

In this quadrant, it is important to map who is mainly impacted by the problem in discovery.

The goal is to understand the profiles/personas that are impacted and the impacts they suffer in their routine.

The reason to learn more is that they are potential product customers besides being potential contacts to learn more about the problems and contexts.

This information is relevant to the Product Driven pillar.

# Business Impact: Risk Costs

The Risk Cost For The Business quadrant is where you map why it is a problem for the business. This must be mapped regardless of whether it is a challenge that generates issues for the business or a concrete problem.

Map all the characteristics related to risks at any aspect (financial, reputation, etc.) that the business suffers and the consequences of it for the short, mid, and long term.

Another important aspect to identify is the willingness of the business to solve that. This information is especially relevant for strategy in the Product Driven pillar. See Figure 7-3.

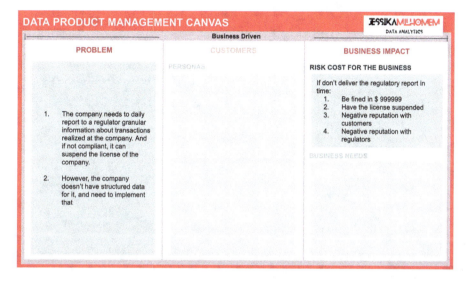

***Figure 7-3.*** *Data Product Management Canvas: Risk Cost Sample*

# Business Impact: Business Needs

Finally, for a holistic vision of the scenario, it's important to understand the expected results once a solution is designed for it. It includes all kinds of expectations (financial, reputational, etc.).

You may have some ideas about it. At least you should. Anyway, even with that in mind, you should also understand the expectations of the business as an output from the product you are creating.

Map all the expectations, and besides the potential customers, who else are interested in it: the potential stakeholders.

This information is important for the strategy in the Product Driven pillar. See Figure 7-4.

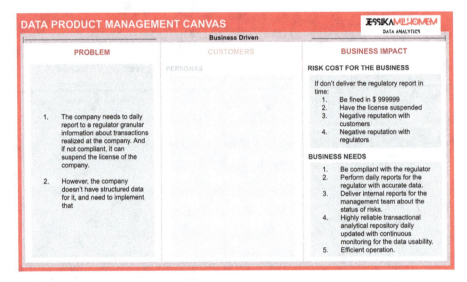

**Figure 7-4.** *Data Product Management Canvas: Business needs Sample*

These four blocks and five suggested questions are the most important to mapping the answers. This is because you will use it to consolidate the proposal for the project. Besides, it is equally important to identify potential topics to go deeper into as soon as you need them during the discovery phase, which also includes the Solution quadrant of the Data Product Management Canvas.

In the end, you'll be able to do the following.

- Validate that you do have a problem that needs to be addressed.

- Have a holistic vision of the scenario. Determine if you should keep investing in it or pivot (learn from the scenario, but give up the path you have been following so far).

- Understand the characteristics of the problem.

- Start understanding the potential customers.

- Map the direct business impact that this problem is generating.

- Have a glimpse of what you need to keep discovering during the solution phase.

- Have a direct, succinct, and sufficient initial conversation with the potential sponsor of the project.

## Business Strategy: Why, What, and How

Besides the direct business context related to the problem, I recommend you also understand the organization's business mission, vision, and goals.

This understanding is a beneficial complementary context because it enables you to recognize more perspectives about the impacts of the problem on the business and the potential desires of the customers.

Many companies work with Objective Key Results (OKRs) to measure the results of enterprise-level initiatives. It's also valuable information to consider during the consolidation of knowledge to plan the project proposal, mainly to design the strategy and work with the Product Driven pillar's quadrants.

# 7.2.  The Target Audience

## Customers: Persona

A persona is how others perceive someone's character. In product management, it's a fictitious personality with characteristics, roles, and behavior that represents the classic user of a product.

They help with the product's design and evolution by considering the personas' needs and expectations.

Now that you know what a persona is, let's return to the framework and process to finalize a discovery related to the Business Driven pillar!

It's the second phase of the Customer Persona quadrant, which is the time to map more characteristics of potential customers.

## Role and Characteristics

The first thing that is important to map about data products is the roles of the customers.

Depending on the organization context, it won't be enough, but it's the first stage.

Here, the goal is to map what the role title is and what their characteristics are. The following are some examples.

- Is there more than one profile?

- If so, is one profile more technical than the others, or are they more user-consumer?

- What type of data governance profile do they have: data owner, data steward, data creator, data user, data consumer, or application owner?

- Should any additional characteristics be considered? Should business context be included here?

This doesn't necessarily mean that the role defines the whole perspective of a person in a company.

Many companies define the titles but don't have a pattern of scopes, for instance. It's a cultural issue that I won't touch on in this book, but it's important to consider. It is important to consider this when marketing products.

In summary, each problem scenario needs to be evaluated considering its own evaluation necessities.

Although I'm sharing many best practices and guidelines for each framework architecture I designed, there's no silver bullet.

To create a strategy, it's necessary to consider the full context, where you must join and apply your experience, strategic vision, and skills. Without it, it's not possible to move on.

## What They Care About and Why

To have a holistic vision of your customers, besides their characteristics, you need to map their behavior.

Thus, considering everything you mapped, you must understand why they would use a product to solve this problem. What should you learn to have clarity about it?

Moreover, once they use it, map what you think they would prefer and what they would despise. This information is important to the next phases when you plan and design the relevant features for the customers.

For this phase, it's important that you understand what they care about and why, by studying their behavior (See Figure 7-5).

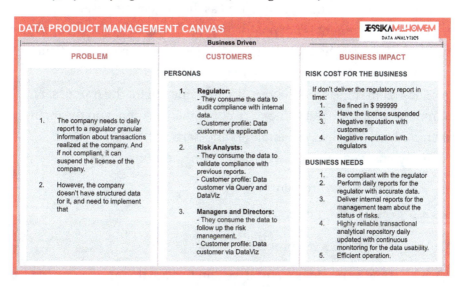

**Figure 7-5.** *Sample of the Business Driven pillar*

# 7.3.  Consolidating Issues and Expectations

The last step of the Business Driven pillar is consolidating the knowledge generated to achieve two main goals.

- Start working on the product-driven phase.

- Create the proposal for the project to align with the potential sponsor.

## Consolidated Knowledge: The Product-Driven Phase

Consolidating knowledge is necessary to ensure the start of work is in the product-driven phase especially. But also to create the proposal for the product design.

In practice, you will write two main documents: one for internal purposes, where you will detail all the information necessary for the product strategy. The other one is used as the project proposal.

The first document is not mandatory to be created by the data product manager. However, their knowledge of it must be fully considered during the strategy phase, as discussed in the next chapter.

The second one is mandatory. It's fundamental for the proposal and communication with the company's upper management and the main interested ones, such as stakeholders and partners.

In general, you must answer these main questions.

- What is the problem?

- Why is it a problem?

- For whom is it a problem?

- How much does it cost if we don't do anything?

- What are the expected results if we work on resolving it?

The difference between them is that for the product design, the detailed information, including technical details mapped while studying the business, is also relevant and must be considered, as well as each business process and context. For the proposal document, you must consider the high-level perspective and provide the information necessary for the project investment.

# Create a Project Proposal

Before the whole product management process, one important responsibility of the data product manager is to "sell the design and implementation of the data product and its potential" first to the organization.

Contrary to what usually happens with other technical products, executives usually do not know the data world and fundamentals. Managers (from different levels) are usually hired to work with data but do not consider any the strategic vision for it. Even the senior ones, who should be especially responsible for it. It cyclely reinforces the common behavior of executives not involve these data leaders on core business strategies. Not since the beginning.

It occurs due to a bunch of reasons, but mainly because the market is unaware of the data world context and are more used to "final business solutions."

Therefore, it's not uncommon for companies to have or hire data managers at companies that do not perform this important responsibility. Besides, in many scenarios, this is due to the managers' lack of knowledge and proactivity.

There is still a journey to evolve the strategic vision and fundamental knowledge for non-data and data executives and leaders. Clarity about data products, including their importance, is fundamental to changing the behavior in the market.

In short, data product managers must study their market, customers, and issues, design the proposal, and sell it before performing its life-cycle management.

Let's return to the section phase, which focuses on designing the project proposal, presenting, and negotiating with the potential sponsor. Also, it is important to highlight that even if you are focused on an external market, you'll start with the internal stakeholders.

Obviously, you'll follow the company's culture approach for the whole communication and negotiation process and respect the responsibilities of each upper-management role. Regardless, you are responsible for creating this concrete proposal and ensuring it flows as necessary for its evolution. And the material you'll create will be a very relevant piece of strategy for this work.

The material must contain the information necessary for a decision taken regarding the project. It must contain the following information.

- **The problem**: What it is, why it is a problem, and who is most impacted

- **The project goal**: What will be solved, when, and the expected results

- **Investments for the project**: What is necessary for the project: financial and time investments

For the investment aspects, clarify the following.

- **Roadmap**: Milestones and deliverables timing

- **Team**: Profiles and scopes

- **Technology**: Any technology prerequisites

- **Communication plan**: When and how to communicate with sponsors, stakeholders, peers, and customers during the project and about it

- **Financial investment**: The total financial investment and results of it delivered by the product during the project.

Consider the culture of the company when sharing this information. For instance, the financial investment context is usually shared just with sponsors. However, it depends on the communication culture of the organization. See Table 7-1 to consolidate the understanding about the two documents and the differences between them.

*Table 7-1.*  *The Business-Driven Phase*

| | | |
|---|---|---|
| **GOALS** | Leverage the detailed knowledge for the product-driven phase. | Share and negotiate with the Sponsor and stakeholders (if applicable) a concrete plan for effective solutions to the problems. |
| **WHAT IS TO BE DONE?** | Focus on the short-, mid-, and long-term plan.<br>• Short term: Implement the delimited solution for the problem.<br>• Mid/long term: Design the strategy for the product. | Focus on the plan to implement the solution for the short term, mentioning the high-level perspective about the mid/long term vision (the product you'll design). |
| **HOW TO DO IT?** | Although it is not mandatory, ideally, through a document. Considering the full knowledge (business and eventual technical details) in the product-driven phase is mandatory. | Through a document for a formal and detailed description.<br>It can follow the company's culture, although my recommendation is to be succinct, but with complete context about the problem, reason for it, necessary approach, investment from the organization, and expected deliverables and results. |

# 7.4. Summary

Let's recap the main takeaways of this chapter.

- **Business Driven pillar**: The steps to empathize with the customers by understanding the business context, problems, and related details. And define what will be solved after consolidating the knowledge and delimiting the issues to be solved.

- **The following are the main questions you should use to map the problems.**

  - What is the problem or challenge?

  - Who is mainly impacted by it?

  - How is it a problem or challenge for the business?

  - How much does it cost to not do anything?

  - What are the expected results if we work on resolving it?

- **Customer persona**: It's fundamental that you understand your customers well to design solutions that will support them. For this reason, designing a customer persona is very helpful.

- **Consolidation of your discoveries for the Business Driven pillar**: It's important for two main purposes: to be leveraged for the product design and implementation, that will be covered on the Data Product Management Canvas' Product Driven pillar, and is also important for the project proposal and presentation you'll implement for the sponsor.

# Summary

In this chapter we talked about each quadrant of the Business Driven pillar of the Data Product Management Canvas and how to use them to discover everything necessary and relevant for the business. It's fundamental to create the proposal for the project and the starting point for the product design.

In the next chapter, you'll learn about quadrants from the Product Driven pillar and how to use them for the exploring phase of data product management.

# CHAPTER 8

# Designing the Data Product: Exploring Phase

This chapter talks about the exploring phase.

I'll cover the Product Driven pillar's quadrants related to strategy and solutions. Both are related to the Data Product Management Canvas that I created for this phase (see Figure 8-1).

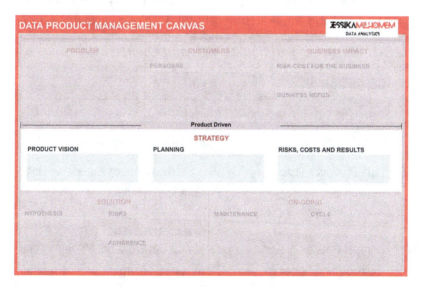

***Figure 8-1.*** *The Data Product Management Canvas: strategy block*

© Jessika Milhomem 2025
J. Milhomem, *Data Product Management in the AI Age*,
https://doi.org/10.1007/979-8-8688-1315-3_8

Let's start with the Strategy quadrant by discussing the product vision, planning, risks, and costs.

After the Strategy quadrant, I'll discuss the Solution quadrant by focusing on its idealization until its prototype.

# 8.1.  Strategy Design

Strategy is critical and fundamental to product design and evolution.

It just makes sense to do it after the approval of the project proposal and defined definitions with the sponsor.

## Business Strategy: Why, What, and How

Regardless of its segment, any organization must follow one strategy to move forward. At least, they should. Otherwise, the failed path has more potential than on the opposite side: the desired success.

For the creation of the strategy, ideally, it must follow a top-down vision that is performed and achieved from a bottom-up perspective (see Table 8-1).

***Table 8-1.*** *Strategy Workflow Design and Ownership*

|  | DEFINITION OF | DEFINED BY |
|---|---|---|
| **WHY** | Company's Vision, Mission and Strategy | C-level |
|  |  | (C-level and vice presidents) |
| **WHAT** | Data Product's Vision, Mission and Strategy | High Senior Management and Sr Data Product Management |
|  |  | (Directors and Managers) |
| **HOW** | Product Operationalization (Solution and Ongoing) | Sr-Mid-Jr Data Product Managers and Development Team |
|  |  | (Managers and Leaders) |

I like the approach Simon Sinek discussed in his book *Start with Why* (Portfolio, 2009). To engage people to achieve something, they need to feel part of it. They need to understand the reasons. They need to understand the whys. And I include myself in that context, especially as a result-oriented person.

Then, before anything, it's fundamental to define the company's vision, mission, and strategy in an organization. The founders, C-level, and vice presidents commonly define this definition.

I won't go deeper into this topic because it's not the book's focus, and considering the main subject covered, these definitions are already set up in the organization.

However, it's critical you know the company's goals and the reasons behind them.

The next part of the organization's strategy must define what will be done to achieve the enterprise's vision and goals, such as processes and product designs. When visualized from a company-level perspective, it may be tactical. However, we know that there are strategic responsibilities, especially considering product definitions.

This phase applies to any aspect of the business, for direct and indirect goals, usually the data analytics plans configured in a company. Senior management and product managers usually define it. In our context, by data product managers.

For that reason, the focus of this chapter is mainly related to this phase.

The final step is related to how the organization's vision will be applied. It's time to operationalize the product vision and processes. Parallel to our focus, it's time to operationalize the data product.

Data product managers and the development team perform that phase. For that reason, it is covered in this book with sequential prioritization.

# Data Product Management and Strategy Design

During strategy design, you define the product vision, design the short-, mid-, and long-term planning, and finally, map and plan the mitigation of the risks and costs related to it (see Figure 8-1).

This work is not only done at the start of a product project but as a recurring review and redesign of the strategy. For success, the cycle needs to be continuous.

After reading the "Product Vision" section of this chapter, you can use the Data Product Management Canvas to design and visualize your strategy.

## Product Vision

The first step in the strategy phase is to define the product's vision.

This process involves the clarity and consolidation of the vision of the Business Driven pillar. In other words, you must have clarity about the problems, risks, and costs for the business derived from these problems, the potential and target customers, their needs and expectations, and complete clarity about the business needs and expectations.

With everything mapped, you can start the product vision design. You must consider the long-term vision for the product. You must realize the target. Think outside the box and be ambitious, but know what your potential customers are facing.

### Define the Mission of the Product

You may have mapped many problems with many different aspects.

What do you want to solve with the product? That's the main question you must consider and answer.

This question is your guide!

## Analyze the Concurrency

As soon as you have clarity about the problems you need to solve regarding the product, you need to map and understand other solutions in the market—externally and internally.

Besides clarity about the market target, it's important to consider the organization's culture, especially if your product is focused on internal solutions. You need to understand if your concurrency is internal, or external, or both of them. Evaluate what exactly they address, and how. What are the values that they are able to deliver, the direct and indirect costs of it, what are their strengths, their weaknesses, and all aspects of the product. It's important to have this understanding, because all this information will support you on designing your whole strategy. Answering questions as:

- Do you really have concurrency or is this market new?

- How mature is this product?

- Would it solve the problems the business face completely?

- Should you leverage any existing solution or create something from scratch?

In short, you need to evaluate all the aspects and think about what will be your strategy.

After getting this information through studies and definitions, it is time to design the high-level product functionalities of it.

## Product Functionalities and Diagrams

You have the complete vision for the problem now, but it must also be clear for the stakeholders and customers.

For this, you need to create a diagram showing product design, its features, its main components, how each component interacts with each other, and so on.

Suppose you are going to work with the Golden Data Platform. You'll use the design and adapt the modules based on your reality and necessities.

Another important vision design to clearly define is how the customers will use these products.

That's when you use the personas you designed during the Business Driven pillar phase.

For this, although not mandatory, I highly recommend you create a diagram showing the possible use cases of usage by all of the personas you mapped. This diagram must connect the customer to the problem faced and the product's usability to address the specific scenario. There's no set recipe for creating this design, as you can design any kind of products as we discussed so far. But here are some suggestions you can try: Flowchart, Use Case Diagram, Data Flow Diagram, any other common diagram, including, the creation of a new one. Again, totally up-to-you decide the best tool to illustrate it. The most important is to create some visual material to give clarity about it, besides only descriptive information.

It must contain all the scenarios the users may face and need to use the product to address the business problem.

## Product Identity

It's time to give identity to your data product!

The first thing to do is define its name, regardless of the data product type.

Ideally, the name should refer to the vision and mission of the product. Anyway, it's important to have a reference for it.

It's also important that it have a face.

**Tip**   There's no hard rule for defining a product name, but consider it based on the vision and mission of the product. Simple but original enough to be easily associated and remembered. The visual identification must fit with the name and also be original.

For products created for internal purposes, if the organization allows it, it can also be a parody by referring to anything from pop culture, as long as it relates to the product's purpose.

Of course, it'll make sense just for some types of products of data:

- For any Data Product, and

- For algorithm, decision support and automated decision-making Data as Products.

Data as a Products as raw data and derived data wouldn't have images for identification because they are super-specific, and having just the name is enough for their identification.

## Consolidating Requirements

You have a vision for your business's dream data product.

Prepare yourself and enrich the vision for the product with a more pragmatic plan, not just for the short term but also for the middle and long term.

To move forward to the next step, by designing and planning for milestones, it's fundamental that you have clarity about the features you need to implement. You need to map all the issues the product is supposed to address.

Besides the issues you mapped to design the project proposal, map the full potential issues the product has to address and its requirements.

It's important to consider the characteristics of the product you're designing: data as a product or data product.

# Agnostic Business and Technical Requirements

Regardless of the product, there are some questions you must consider for your evaluation.

- Are you working with a product that should consider geo-globalized requirements?

- If so, what must be agnostic for any geo-agnostic requirements?

- Besides the hard definitions (a.k.a. mandatory or nonnegotiable), are there common definitions in different locations?

- Are there mandatory definitions that must follow local definitions?

- Are there technological considerations?

# Agnostic Technical Requirements

Besides that, another critical point is related to the infrastructure of your product.

If you are working with data as a product, you might have some clarity about the Golden Data Platform after following the path shared in Chapter 5. Anyway, it's important you design the strategy or plan related to this topic.

I suggest you return to Chapter 5, especially if you are working with data as a product, to recap some points and consider them in your plan.

The following are some questions for you to evaluate.

- What is necessary for the product in terms of infrastructure? What kind of features?

- Will it be on-premise or cloud?

- Will we be responsible for the infrastructure on a daily basis, or will a third party?

- Should we consider a budget for an infrastructure platform? What about staffing?

See Table 8-2. Again, I'm only talking about some questions. You must map the whole context you have and should consider to keep evolving your plan.

*Table 8-2.* *Requirement Types for Data Products*

| REQUIREMENT TYPE | WHAT TO CONSIDER |
|---|---|
| Agnostic Business | All business requirements are related to the business's main purposes: financial, customer management, operation, legal and regulations, and so on. For this, always consider the 5Ws: What, Why, When, Who, Where. |
| Agnostic Technical Requirements | All the technical requirements to ensure product execution are as follows: Infrastructure, technologies involved, suppliers (internal or external), legal and compliance, and so on. For this, always consider the 5Ws: What, Why, When, Who, Where. |
| Data as a Product | All the requirements necessary for your product, considering the type of data as a product you are managing (see Chapter 1). For this, always consider the 5Ws: What, Why, When, Who, Where. I recommend you work with the Golden Data Platform (see Chapter 5). |
| Data Product | All the requirements necessary for your product, considering the type of data product you are managing (see Chapter 1). For this, always consider the 5Ws: What, Why, When, Who, Where. |

# Data Product and Data as a Product Requirements

As discussed in Chapter 1, there are a lot of categories for each one of them.

For that reason, it's important that you consider the characteristics of each one to map potential issues and requirements. At the end of this stage, it's expected that you have a clear definition of the Product vision. See Figure 8-2.

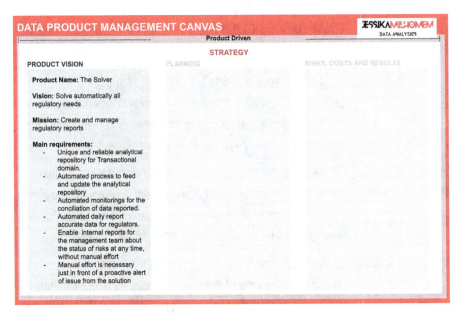

***Figure 8-2.*** *The Data Product Management Canvas - Product Vision Sample*

# Planning

With the product vision concluded, it's time to plan its strategy.

For that, you must first have clarity about the vision and mission of the organization. Then, clarity about the vision for the product created.

The planning step helps you understand the business strategy, define the product and project management strategy, and the measurement process.

For guidance, let's keep using the Data Product Management Canvas. In the Planning quadrant, you organize the order of milestones to be achieved by ideal prioritization, also pointing out the timeframe for their availability.

Besides, you'll define the project management strategy and the measurement approach.

At this phase, you compose the strategy rationale with a macro vision of the plan. Here, you'll set up the short-, mid-, and long-term vision.

It is important to give your customers, shareholders, and stakeholders more clarity about the vision and expected pragmatic deliverables compared to their investment.

Besides, it's fundamental to guide the team on a clear strategy.

Won't it be better to do it for the short term since we'll learn and discover lots of things that will impact our plan?

I understand that many professionals think and perform like that. I'd say the majority work and think like that. At least, that's the experience I have had for almost 20 years.

However, I completely disagree with it! And I will tell you the reasons!

Without a plan, there's no direction. Although the future is not completely certain, it's important to have a mid- to long-term plan to guide you in this direction. Here, I'm talking about six months to two years at least.

It's intrinsic that the plan may have adjustments and changes during the mid/long term period. It's okay. It's expected and natural.

The most updated and clear plan mitigates possible risks for the short term (one to three months, at a maximum of six months).

Then, it's critical that you create a general plan for the product by touching the short/mid/long term, even knowing it will have changes.

Giving more confidence to your stakeholders, shareholders, and customers is necessary.

And more than that, besides the product vision, it'll be a shining light for you and your team to follow while discovering and innovating.

It is important to keep reviewing and updating it to ensure the success criteria defined for the project achievement.

## Product's Functionalities and Roadmap

With the features mapped for the product, it's the moment to plan when they will be implemented.

Of course, to define the roadmap, it's also important to have clarity about the team allocation you need to work on.

Data professionals' roles and responsibilities are covered later in the chapter. For now, let's focus first on the roadmap and necessities of it.

Here, you may face one of the two scenarios.

- A budget for you has already been defined regarding headcount allocation and considering the project proposal you created and approved.

- You can re-evaluate the required headcount allocation by considering the short/mid/long-term plans.

I'd say the majority are in the first scenario. We usually face it and the adjustments, as the second scenario becomes an opportunity as soon as the results are presented in the first one.

Thus, here is the moment to focus on the milestones deliverables.

- **Short term**: This is ideally the approved project proposal you presented. This strategy focuses on addressing the most critical issues for the business that bring impactful results.

- **Mid term**: The plan is to consider the potential and desired milestones to be achieved to keep developing and evolving the product as soon as the first main deliverables are done. Think about what you'd like to do as soon as the project is concluded.

- **Long term**: This is the vision for the long term. Again, it may change during the journey, but having a plan is fundamental, including awareness of potential adaptations and changes.

Always consider the Business Driven pillars as the foundation for the whole strategy.

## Data Team Roles and Responsibilities

Regardless of the scenario you may face regarding the composition of the team, whether the budget is defined or not, you must know what kind of work is necessary.

Knowledge about roles and responsibilities is fundamental because it can directly impact the roadmap's execution and success.

In the past, we heard about "unicorn professionals." But it is now clear that it is not feasible to compound this due to the accumulation of expertise to perform everything, not just in knowledge but especially in practical work.

The reality is that there wasn't enough clarity about the potential work to be done; we were learning about it.

Meanwhile, the market has kept evolving, and the necessities have become more complex.

Although there were changes, I'd say that there are three main areas to be considered for data teams: business analytics, artificial intelligence (which the market points out just as data science), and data engineering (which the market points out as software engineering, but I consider data, as it may fundamentally change the business focus of the role).

Besides, of course, the necessary data governance responsibilities are here.

The main difference is that these big areas require specific responsibilities that become specializations. Thus, specialists with clear scopes are mapped as data team members.

In general, there are three big domains.

- **Business**: Business domains are responsible for the operationalization of the business responsibilities in their areas, besides the analytics

- **Data engineering**: Responsible for the infrastructure, data storage, and computing of analytical data for business purposes

- **Artificial intelligence**: Responsible for designing and operationalizing the algorithms for advanced analytics

- **In addition to data governance**: The organization-level team is responsible for defining the policies for the governance of related data

Each area has one or more specialist roles (see Figure 8-3).

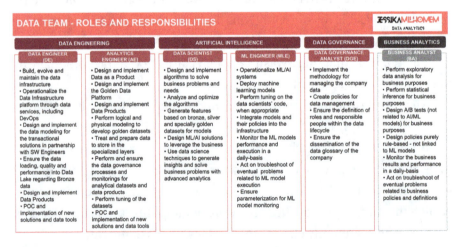

***Figure 8-3.***  *Data team roles and responsibilities*

## Data Engineering

- **Data engineers**, also called data platform engineers, are responsible for the data infrastructure platform and perform the data steward responsibilities of transactional repositories.

- **Analytics engineers** are responsible for implementing the Golden Data Platform, data as a product and performing the data steward responsibilities of building data pipelines.

## Artificial Intelligence

- **Data scientists** are responsible for designing and implementing algorithms to solve business issues using advanced analytics techniques. They are also data stewards for data science responsibilities.

- **Machine learning engineer**: responsible for operationalizing ML/AI systems, from deployment until its monitoring and ongoing. They are also data stewards for ML/AI responsibilities.

## Business Analytics

- **Business analysts** perform business responsibilities related to the actual operational, and analytics to generate insights and monitoring for business sustainability.

You may have faced some mixed profiles in your professional experience. This happens in the market mainly because some specifications are not common for some roles.

While I recognize that there is no widely accepted definition in the market for this, the definition I have chosen to adopt in this book is the one that I consider most appropriate for the context, as it is the one I believe to be the most accurate.

Another reason is related to the amplitude of scope discussed in Chapter 5 (see Table 5-2 and Table 5-3).

***Table 8-3.*** *Measuring Results of Data Products*

| BUSINESS PERSPECTIVE | MAIN QUESTIONS TO DEFINE OBJECTIVES | SOME EXAMPLES OF MEASURES FOR OBJECTIVES DEFINED |
|---|---|---|
| **Customer Perspective** | To achieve our vision, how should our customers see us? | • CSAT (customer satisfaction score)<br>• NPS (net promoter score)<br>• CES (customer effort score)<br>• Customer churn rate<br>• Customer retention rate |
| **Financial Perspective** | To achieve our vision, how do we look to shareholders? | • ROI (return on investment)<br>• ROE (return on equity)<br>• Net profit margin<br>• Gross margin<br>• Business operational cost savings due to the product data<br>• Data costs savings |

*(continued)*

*Table 8-3.* (*continued*)

| BUSINESS PERSPECTIVE | MAIN QUESTIONS TO DEFINE OBJECTIVES | SOME EXAMPLES OF MEASURES FOR OBJECTIVES DEFINED |
|---|---|---|
| **Internal Perspective** | To achieve our vision, what must we excel at? | • Businesses succeed due to data analytics product rate<br>• Data quality rate<br>• Data governance rate<br>• Efficiency increase rate<br>• Cycle time<br>• Actual introduction schedule vs. plan<br>• Time to market |
| **Innovation and Learning Perspective** | How will we sustain our ability to improve and create value to achieve our vision? | • Innovated processes rate<br>• Time to improve issues<br>• Communication efficiency rate among areas |

In addition, size usually impacts it. The tendency is that bigger companies need more specialized professionals, and smaller businesses need more generalists, accordingly to the diversity of use cases and business complex. However, as we have seen the movements in the market, the companies, including the biggest ones are specially interested in having shorter teams to work in squads based on the business domains and needs. For this reason, it's also becoming common the merge of the specializations, consisting in what I'm calling data generalists, although they are specialists (or desired as).

There are three types of data generalist professionals (see Figure 8-4).

- **Analytics Engineering generalists**, by accumulating analytics engineers, data engineers, and business analysts responsibilities

- **Data Engineering generalists**, by accumulating analytics engineers, data engineers, machine learning engineers, and data scientists responsibilities

- **Data Science generalists**, by accumulating business analysts' and data scientists' responsibilities

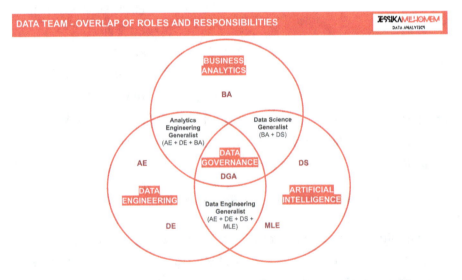

*Figure 8-4.* Data team: overlap of roles and responsibilities

---

**Note** Eventually, you may face even more overlap when matching all specializations, such as, merging the Data Engineering generalists with the Data Science Generalists. It may occur especially in startup companies with a small staff and many responsibilities. In short, overlap may be common for small companies.

---

Again, defining the scope of the organization must consider many aspects.

I see value in both approaches. However, with the evolution of data and AI technologies, summed with the necessities that are emerging and market behavior, I see that it may be the path seen more in the mid- and long-term for data analytics professionals in the market.

## Measurement Definitions

Measuring product results, including success metrics, is very important. After all, without it, it's not possible to map any achievements. Besides the DMAIC approach discussed in Chapter 1, I also like using a balanced scorecard.

A balanced scorecard (BSC) is a strategic management tool that Robert Kaplan and David Norton created. They introduced it in 1992 by a paper published in the *Harvard Business Review*.[1]

In 1992, they wrote a book to further explain their ideas regarding this strategy tool.

The main difference of the balanced scorecard was the innovative rationale of not just considering the financial aspects to measure the growth and success of a business or an organization but also other relevant aspects. It comprises financial, customer, internal business processes, and learning, and growth.

More than that, the BSC translates the business mission and strategy into objectives and initiatives in a decomposed manner, using the four perspectives.

It's a great tool for defining objectives and measuring KPIs (see Table 8-3).

Of course, it must be driven based on the business context you are inserted and the problems you're addressing with the product you're designing.

---

[1] Robert S. Kaplan and David P. Norton (1992), *Harvard Business Review*, "The Balanced Scorecard—Measures that Drive." Available at https://hbr.org/1992/01/the-balanced-scorecard-measures-that-drive-performance-2

Thus, as soon as you have clarity about it, define the objectives and the KPIs to use as a reference, as shown in Table 8-3.

You should design the measurement definitions strategy not just for the product vision but also for the plan for the exercise period of work you have or will have.

The usage of the BSC is not needed for any type of analytics because the metrics and KPIs you define must be efficient. The amount of it is also part of your evaluation. Thus, you must design your strategy accordingly with the purpose of the measurement you'll do.

What I can say is that, overall, this approach is especially relevant to measuring the results of the business and the product's results for the business. But it's not necessarily a demand for the strategy design of your work period.

Possibly, your company works with objective key results (OKRs). If it doesn't use this management tool but uses another one, it's okay. The important thing is to have one efficient tool for the same purpose: To manage the business with a unique vision, target, and clear strategy. It is important that you leverage it to create your strategy. For your product!

In this book, OKRs are the tool for that purpose. To recap what this management tool is see Chapter 1.

To design the strategy goal of the exercise period of work with the product, especially focused on the short term, you should use the cascaded strategy described in Table 8-1.

In other words, by relating the OKRs of the company, you'll design the OKRs for your product in a way that supports the strategic organization's OKRs through the targets related to the product you're designing.

This approach is fundamental for the canvas to put the product in production pragmatically.

# Project Management Approach

The strategy plan is designed. Let's go deeper into the tactical plan for the idealization of the solution.

Regardless of the type of product, as mentioned in a 2002 *Harvard Business Review* article,[2] it's fundamental to apply key product development principles, including agility and iterability. In other words, dynamic and faster solutions. It, of course, also refers to the approach used to manage the project.

Project management is also part of data product management. However, it can be delegated. It can be performed by one specific professional specialized in it, not necessarily the data product manager.

Another way to do it is based on the proposal I've been sharing in this book, and it is my favorite: have the project management performed by data product managers but by different levels of detail. For this approach, data managers responsible for each scope are required. To better understand it, let's talk about the tools for it.

The project contains two main roadmaps.

- **Macro roadmap**: It is a holistic roadmap with defined main features and milestones. It does not consider just the short term but also the mid/long term, which can be shared or not, according to the audience and strategy of communication. This roadmap only communicates with sponsors, stakeholders, and customers. But it's also used for vision communication and alignment with the technical team.

---

[2] Jedd Davis, Dave Nussbaum, and Kevin Troyanos (2020), "Approach Your Data with a Product Mindset," *Harvard Business Review*. Available at https://hbr.org/2020/05/approach-your-data-with-a-product-mindset

- **Detailed roadmap**: This roadmap contains a detailed plan. It is a zoom-in of the macro roadmap. It contains each incremental (small deliverables that bring value to the product and business) planned. The detailed roadmap is mainly used for internal communication and is the foundation for the operational execution and planning of the team, including integration with other areas.

Both roadmaps are related and dependent, but they have different purposes and ways to be managed.

- **Macro roadmap**: It is created and managed by the product owner, the data product manager. It is created from an internal perspective to share the product's vision and the strategic path to follow with the team. Of course, I see it must include collaboration from the team in terms of onboarding and alignment of effort. However, it is mainly designed by the data product manager. After that, it will be updated just after a strategic vision is adjusted (the main reason for that). Or when it's necessary to adapt to any specific unexpected scenario during the journey. Then, it'll follow and be updated with the timeframe from the detailed roadmap. The product owner is the final decision maker for prioritizations and initiatives to follow.

- **The detailed roadmap** is managed mainly by the junior data product manager and is designed with all team members, especially the technical team. The reason for that is this roadmap must contain a detailed plan to ensure the detailed incremental, which requires

profound participation from the technical team. After
all, although the data product manager is the product
owner, everybody also owns it. They must participate in
the process.

It may appear as more work, and it can be at the beginning; however,
this organization and strategy speeds the team's work process, adds clear
strategy visibility, and increases agility, which is one of the main reasons.
The second one is to ensure the integration and collaboration of all levels
of execution of the work.

As I usually say to my team, "To ensure success in the work, the team
must work as cells to compound an organ. It must perform as a network,
with everybody integrated, not siloed. If one piece fails, it'll impact the
whole system. In the same way, if it works healthily, the capacity and
probabilities to succeed increase."

After defining the responsibility for the project management and the
process, the next step is defining the methodology approach.

Again, there's no unique rule or silver bullet. Each can be performed
from different perspectives. You must find the best way for you.

The way I like to perform and suggest you apply is to use the Scrum
methodology for it while you are working with the project. You may have
been thinking: "But you have just mentioned about creating roadmaps,
and one with detailed rationale. It seems to be a waterfall approach,
isn't it?".

You are not wrong about your rationale. Creating a roadmap plan is
similar to the waterfall approach, but it doesn't mean you cannot use the
agile approach.

I like to use both of them. The reason? Adaptability solves what is
necessary for what I usually need to attend to my demands: business
needs and people management needs.

First, let's talk about the business needs. Let me know one thing: Would you invest your money and your time in a project performed by other people, that you don't know what will solve with more pragmatic clarity and when you have each macro deliverables? If you are not super rich and has a huge availability to spend money, I bet your answer is no! Now, if you do it, imagine high-management professionals. They are responsible for the health of the company and many jobs maintenance. So, the short and obvious answer is: we need to have and give clarity about what we'll perform, about the management. And as a Data Product Manager you must do it. Period.

The second necessity in my management approach is that my team must be part of the work and have the flexibility to experiment, test, and apply while generating value. We must do it together. For this reason, I like the agile approach. Besides, I do appreciate the teamwork. I usually say to my team that I do perform very well alone. However, I enjoy the potential of teamwork. That's one of the reasons I decided on the management path. So, I manage that way. Again, there's no silver bullet or wrong and correct approach. However, it's best to follow a recommended best practice approach. Each professional must follow their own. Applying their values and visions.

For these reasons, I like to use Scrum. It offers the tools necessary to attend to my needs.

You must define the whole project management approach for your product development and the processes you will use. At the end of this stage, it's expected that you have a clear definition of the Planning vision and it clearly mapped in the Data Product Management Canvas. See Figure 8-5.

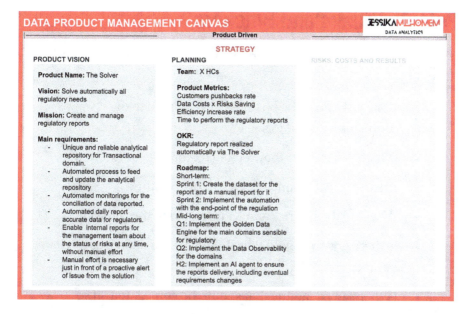

*Figure 8-5.* *The Data Product Management Canvas - Planning Vision Sample*

# Risks, Costs and Results

The last quadrant of the strategy phase is related to the business risks and costs. This phase maps the potential risks and blockers and explains how to solve them.

With the macro plan created, map the potential issues here from internal and external perspectives. The goal of this phase is not to go deeper into all of the details but to a holistic perspective.

Remember, you're designing your strategy. The detailed scenarios will be evaluated at the appropriate time.

# Potential Risks and Blockers

It's important to map at least the following questions to understand the risks.

- What risks does the business have related to the problems we need to solve?

- Which process are they related to: internal or external?

- Are they legal, financial, reputation, or other?

- What are the metrics and KPIs related to these risks?

- Is there any threshold that should be considered as a cut-off?

- Is there any deadline that should be considered? If so, what is the timeframe?

- What will happen if the deadline is not met?

- Who are the entities involved with these risks?

- Are there any internal teams that are also interested in these topics?

- What kind of blockers could appear?

- Is there any potential blocker that would make the project unfeasible?

The main goal is to address all perspectives related to the risks.

It must be considered not only for the short term but also for the mid term and the long term. The evaluation must be done for all milestones mapped for achievement.

Of course, the resolution's focus is mainly on the short term. However, it's important to understand the other perspectives to be prepared for any scenario requiring longer projects for resolution. Even more complex are resolutions that require partnerships or prerequisites that require effort from others besides you.

# Handling Risks and Blockers

The mapping of the risks and blockers is important, as is their mitigation plan.

In the same way, the main focus must be on the short term. However, having a general perspective for the mid/long term is fundamental.

The following are some questions to answer.

- What exactly is necessary to solve it? Why?

- Will it be a real resolution or a workaround?

- Would a workaround solve it, or is a final and complete solution necessary?

- How long will it take?

- Will it be enough to solve the problem?

- Is it possible to mitigate or resolve the blocker by ourselves?

- Is a partnership necessary? If yes, who? What needs to be done?

This is a list of some questions to help you map out the design of the plan. The goal is to map possibilities and design a strategy to solve the blockers and risks you identify.

That's when you apply your investigative and strategic skills.

Remember, this process must be as long as necessary for your business needs.

## Expected Results

The same way, it's fundamental to have clear metrics related to the product's planning, it's fundamental to define with clarity the metrics related to the results, considering the needs for short, mid and long-term. For this process, you can use the same rationale described in the Table 8-3.

At the end of this stage, you should have the concluded Product Strategy designed and it clearly mapped in the Data Product Management Canvas. See Figure 8-6.

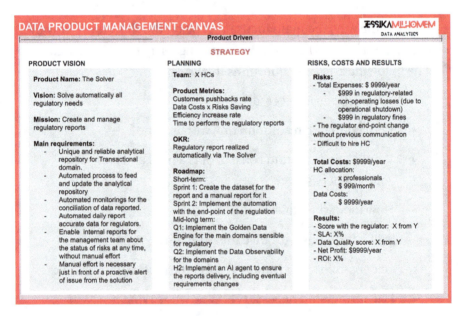

***Figure 8-6.*** *The Data Product Management Canvas - Sample for the Strategy block*

# 8.2.  Idealizing the Solution

Now that you have idealized the product and designed and planned the strategy let's go deeper into the defined short-term plan and put it into practice.

This chapter focuses on creating solutions with tests and prototypes.

To design the solution, let's keep using the Data Product Management Canvas's Product Driven pillar but now focus on the Solution quadrant (see Figure 8-7).

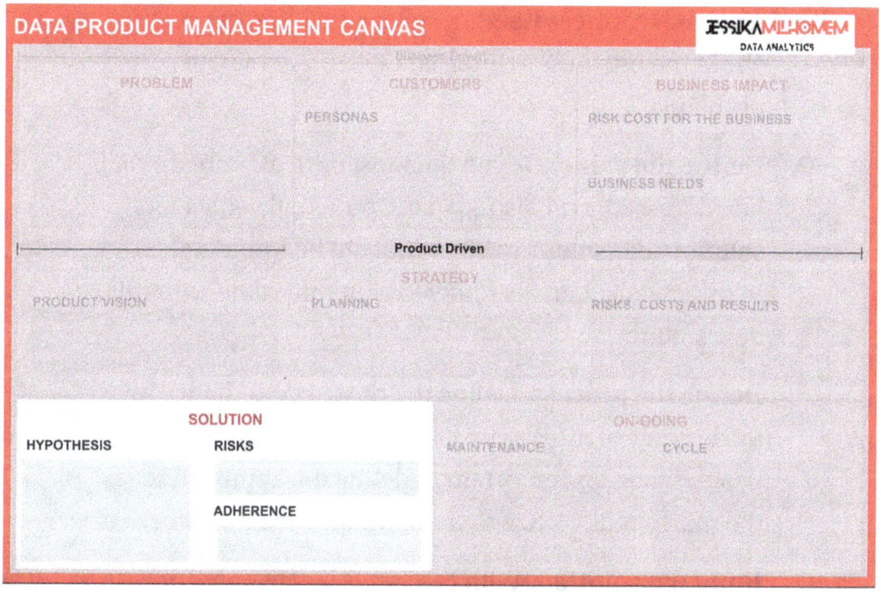

*Figure 8-7.* *Idealizing and building the solution*

After reading and learning all the details described in the following sections, you can use the Data Product Management Canvas to exercise and visualize the hypotheses and risks.

# Minimum Valuable Product

Before digging deeper into methods, it's important to talk about the minimum valuable product (MVP).

Frank Robinson introduced MVP but it became known due to Eric Ries, who wrote *The Lean Startup* in 2011.

The MVP approach is to design and implement a simplified version of a product to validate its value. The goal is to implement the solution with as much reduced time, effort, and resources as possible and validate its value.

The following steps are necessary in MVP.

1. **Map the business needs.** Map all the information about the business's (internal or external market) problem.

2. **Define the value.** Define the most critical problem to be addressed, and also map the potential value of its solution, considering the product to be implemented. In other words, answer how to generate value with this new solution.

3. **Define the process.** Define the product's audience and the journey of the product's user to solve their problem. Besides that, it's important to define the approach to test the market adherence to the solution once it's launched.

4. **Build the solution.** With the necessary features or features prioritized, which must be the minimum necessary but bringing the necessary value, implement it and build the solution.

5. **Launch the MVP.** It's time to launch the MVP for the test group, which was defined before and with the designed strategy.

6. **Learn from the data and feedback.** Finally, it's time to measure the results of the solution and its adherence to the market. Apply BML (build, measure, learn).

Why am I talking about this? MVP must be part of the routine of the data product, and it's a fundamental concept. Moreover, it's the foundation for the Solution quadrant.

We'll use this approach for every round during the product life cycle.

Besides, this concept is fundamental to reinforcing sponsorship support, as its goal is to validate the results faster and with small resources.

If the validation is successful, the project will continue receiving resources. Otherwise, it's necessary to perform another MVP. This process is called pivot.

# Hypothesis

Hypothesis is the first quadrant in the solution.

The goal is to focus on mapping all possibilities to solve the problem.

Remember that we already have the strategy, but now we're focusing on the tactical plan. Thus, the focus is to work in the short term.

Once the problem is tackled and the OKRs for the exercise period are clearly defined in the strategy movement, let's focus on defining the hypothesis.

# Understanding Issues Through Data

To be pragmatic, besides the problem's solution target and the OKR definitions, it's necessary to have data about the issues.

The first step for the hypothesis approach is to investigate the data and get evidence about the issue. It's fundamental that you map and get access to the databases that are related to this issue.

At this phase, you apply analytics for the data using the following approaches.

- **The data product manager applies discoveries from the data** related to the problems faced by the customers that are necessary for designing the hypothesis. Then, share the full rationale with the technical team to work on executing the product tests.

- **The tech team applies discoveries from the data** to understand the problems faced by the customer and designs and delivers evidences results, together with the hypothesis plan through a blueprint[3] by the end of the work, to be validated by the product owner/ product manager.

- **Both approaches are considered during the product journey**, based on the strategy designed for the product development. That's my favorite approach. After all, everybody is part of the team, and all should have the product's ownership mindset, although the strategic final decision is of the product owner/ product manager.

The most important requirement of the approach is the efficiency and impact of the strategy.

Then, take advantage of your professional experience and vision!

# Designing the Hypothesis

With all the evidence related to the issues faced, it's time to design the hypothesis.

The hypothesis describes the best way to technically solve the issue and prioritize it for specific improvement.

Thus, the goal is to state the hypothesis based on the data found and then propose a solution.

---

[3] A blueprint is a document design created with the rationale to solve some specific issue. It is composed with the actual problem scenario, and with the design of the proposed solution.

Together with it, it's important to define the methodology to validate it by considering the data product type. However, regardless, it's important to consider any experimentation.

- Who should be the public selected to validate the product?

- How will the approach be realized? What is the process for the user to experience the product?

- When will the experiment start?

- How long will the experimentation take?

- When do we measure the results?

It's important that you consider all perspectives and needs for the complete evaluation of the hypothesis. Finally, at the end of this stage, you should have clarity about the hypothesis you'll consider and it clearly mapped in the Data Product Management Canvas. See Figure 8-8.

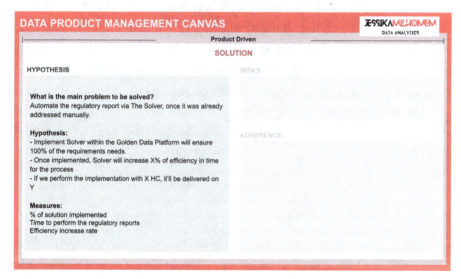

*Figure 8-8.* The Data Product Management Canvas - Hypothesis Sample

# Risks

The risk mapping and mitigation are as relevant as the hypothesis design.

The Risks quadrant is related to mapping all risks, and the migration plan is related to the short-term focus.

Here, you list all the potential risks and blockers related to the business and technical aspects.

For this, you follow the same approach described for the risks and costs for the business that were described in the previous section, focused on the strategy. However, its mapping considers the following characteristics.

- Drill-down the strategic vision, but focus on the prioritized problem and short-term target definition.

- Consider the hypothesis results after it's mapped.

It's also important to consider the relevance and weight of the new risks mapped and evaluate if they must be evaluated and considered in the strategic plan, considering the short, mid, and long term. At the end of this stage, you should have clarity about the risks and have it clearly mapped in the Data Product Management Canvas. See Figure 8-9.

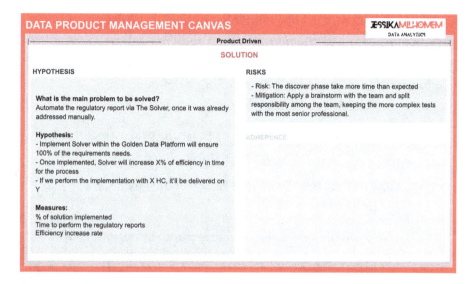

**DATA PRODUCT MANAGEMENT CANVAS**

JESSIKA MILHOMEM
DATA ANALYTICS

Product Driven

SOLUTION

HYPOTHESIS

**What is the main problem to be solved?**
Automate the regulatory report via The Solver, once it was already addressed manually.

**Hypothesis:**
- Implement Solver within the Golden Data Platform will ensure 100% of the requirements needs.
- Once implemented, Solver will increase X% of efficiency in time for the process
- If we perform the implementation with X HC, it'll be delivered on Y

**Measures:**
% of solution implemented
Time to perform the regulatory reports
Efficiency increase rate

RISKS

- Risk: The discover phase take more time than expected
- Mitigation: Apply a brainstorm with the team and split responsibility among the team, keeping the more complex tests with the most senior professional.

ADHERENCE

*Figure 8-9. The Data Product Management Canvas - Sample for Hypothesis and Risks*

# 8.3.  Prototyping the Solution

So far, you have mapped the hypothesis and the risks. Finally, it's time to prototype the solution!

The objective of the prototype is to give more clarity about the purpose to the customer and, with that, be faster on the validation of the product. In this case, with the hypothesis validation.

## Use Case Definition

A good way to do this is by designing the process the user will follow to solve their problem using the product. In short, design a use case for it.

For the use case design, you can talk, write, design a diagram, or merge parts of all of them. It'll depend on the strategy you'll use to validate it. It's not mandatory to document anything. It's just relevant for the full product vision.

For this moment, the important thing is that for the use case itself, you link the customer—the persona of the audience selected for the experimentation definition—to the problem you are addressing and the solution you're designing with the hypothesis defined.

Thus, you'll design the process.

- What is the rationale for the use case, its processes, and flows? It'll guide you in prototyping.

- How will the interface look? How will the customer use the product?

- What will the customer need to see and/or do with your product to get the final result?

# The Prototype

The prototyping should be performed in the simplest way possible, with as simplest approaches as possible to avoid overexertion.

Of course, coding is the best scenario because it can be leveraged afterward for the materialization phase when the product is put into production.

Still, depending on the time and complexity of validation, you can use simpler approaches. As an example, depending on the context of your product's feature, it can be done via drawing on paper to illustrate the customer's journey.

However, when discussing data products, I'd say we have a BFF (best friend forever): The spreadsheets! Yes, the eternal beloved tool by customers.

I know it won't work in scalability in the big data era. However, for samples, to validate an idea, it is a really good option and solution especially for data as a product designs! Of course, it won't be able to address all types of data products, but it might be useful for a variety of them.

Anyway, the point is, don't get hung up on coding only. If necessary, be creative and use other easiest ways, such as tools that make your prototyping easy to validate!

# 8.4. Summary

Let's recap the main takeaways of this chapter.

- **Strategy**: After the Business Driven steps covered in the Data Product Management Canvas, the next fundamental step for the product design is the strategy. It's similar to an organization: without vision, it doesn't make sense and has a huge potential to fail. It's the foundation for the product design and is grounded on the business needs and requirements.

- **Idealize**: When the strategy is designed, idealize the solution and test it by applying data analytics and creativity.

## Summary

In this chapter we discussed each internal quadrant of the Product Driven pillar's Strategy quadrant to design the strategy for the product in the short, middle, and long term. It also covered the Solution quadrant by covering the Hypothesis and Risks internal quadrants related to the idealization of the product.

In the next chapter, you continue to learn about the Product Driven pillar's Solution quadrant but focus on the materialization of the product.

# CHAPTER 9

# Designing the Data Product: Materializing Phase

You learned how to understand a problem by mapping the business needs through the data product management's Business Driven pillar. You learned how to idealize the product by first defining the strategy for the product and exploring the detailed context. You did it through the data product management's Product Driven pillar by covering the Strategic quadrant and the hypothesis and risks portions of the Solution quadrant.

This chapter concludes coverage of the Solution quadrant by discussing the materialization phase of a developing product.

Let's discuss the testing and implementation phases to build the solution.

After reading and learning everything described in this chapter, you can use the Data Product Management Canvas to map and visualize the results for the solution you idealized and tested.

© Jessika Milhomem 2025
J. Milhomem, *Data Product Management in the AI Age*,
https://doi.org/10.1007/979-8-8688-1315-3_9

# 9.1. Testing the Solution

The product is idealized by defining its vision and strategy.

Next, the hypothesis is designed, considering the risks to be managed and mitigated.

The prototype is created, and the group's strategy for testing it is clear.

Then, it's time to test and evaluate the results.

## Adherence

Adherence is the process of evaluating the results of the product.

Let's work on it by following the solution's adherence to the internal quadrant of the Data Product Management Canvas (see Figure 8-7).

It'll happen in two moments: after implementing the prototype and after the product is put into production.

This section focuses on the validation after the prototype, which means that the goal is to map and consolidate the results of the tests applied so far.

For the adherence mapping, it's important to have clarity about the following.

- The method of the validation

- The control group selected for the test

- The metrics that must be evaluated

- The thresholds and definitions for a successful diagnosis

- The thresholds and definitions for a failed diagnosis

- The test period

- Data tracked during the test period

# Evaluation Methods

Quantitative and qualitative information are necessary and important to enrich the analytics during the product's life cycle.

Although both are important, I'd say the qualitative approach has even more relevance for the prototype evaluation because the goal is to have a faster learning curve during its design. And there is nothing better than collecting the information directly from the interested people.

It can be done in different formats; again, it's up to you to define your strategy.

## A Live Group Interview

Based on a sample of selected customer targets, a group of customers is involved in a group interview agenda with a moderator, ideally from one to ten potential customers.

The approach of the agenda is to present the prototype and collect feedback about the product of features implemented.

- The benefits of it are faster answers about the information you need and face-to-face interaction.

- The cons are that the opinions of others can influence the participants, and it isn't possible to map a detailed perspectives of each interviewee.

## A Live User Interview

The user interview is an active discussion with your potential customer. It's a one-on-one conversation.

The approach of the agenda is to present the prototype and collect feedback about the product of features implemented.

- The benefits are the detailed answers to the information you need and the one-on-one interaction.

- The cons are a longer time required to apply this dynamic, the reduced variety of interviewed people, and the fact that a sample of customers is needed for discussion.

## Online User Interview

The user interview is an online feedback form with your potential customer.

You can work with groups of customers or the full database.

The approach to the feedback is to present the prototype somehow (video, demo, trial, etc.) and collect the feedback about the product or feature implemented.

- The benefits are the detailed perspectives of different customer profiles and possibly faster results.

- The cons include the potential reduced volume of answers because it requires more organic engagement of customers.

## Hands-on Usability Testing

Finally, you can get the information from the solution designed if you implement the solution through coding.

The approach to collecting feedback enables the potential customer to use the prototype solution to accomplish the use case defined for validation while being observed.

- The benefit of this approach is the feedback on the actual product, which evaluates the usability behavior of the customer in real time.

- The cons are that it is feasible if the prototype is actually the product coded, which requires more effort and is time-consuming.

# Perspectives to Evaluate

The metrics designed for the test should consider the following perspectives.

## Problem Resolution

The first aspect to consider is related to resolving the business problem. The objective is to evaluate if it's behaving as planned. You and your team should evaluate the solution first. You can use some of the following questions.

- **Use case results**

  - Is the problem effectively solved? If partially, how much? If not, why?

  - Is the resolution relevant to the business need?

- **Features and expectations**

  - Is the feature performing as expected?

## Customer's Perspective

The other aspect that is fundamental to evaluate is the customer's perspective.

The objective is to get feedback on the product's usability and functionalities to solve the customer's problem. You can use some of the following questions.

- **Solution usability**

  - Was the customer behavior expected?

  - What is the positive feedback for the solution?

  - What is the negative feedback for the solution?

- Does it solve potential customers' needs from their perspectives?

- If not, what is not solved and why?

- **Map any missing critical functionalities**: Based on the data analyzed, you may identify that some functionalities are missing and should be considered a priority by requesting a pivot of the plan, or they are improvements that enrich the designed solution. Anyway, this information is relevant. And you should ask yourself the following.

  - What functionalities are missing?

  - Are they hard blockers or improvements?

- **Validation of the group selected**: Based on the feedback, it's also important to evaluate if the selected group of people is appropriate or if it should consider other or additional people for the evaluation.

  - Are the personas involved the appropriate ones, or should they be considered other profiles?

  - What is the impact of the evaluation, and what should be done about it?

The goal of this phase is to have clarity on the result of the product design and be able to answer the main question: "Should we move forward on the implementation of this solution and have the first release of it, or should we pivot the hypothesis plan?"

# 9.2. Implementing the Solution

The solution will be implemented and put into production once the prototype is validated. This process is also called release.

However, you may have heard about different release names. If not yet, you'll learn now! Let's first talk about the release stages because they are necessary for the strategy of the release moment.

## Release Stages

There are six types of software releases, as defined by the software release life cycle (SRLC). They are called release stages. That is because they are part of a cycle (see Table 9-1).

You'll observe that many concepts are related to many contexts discussed in this book.

Although the SRLC is clearly talking about software, which may be more related to data products, it can also be leveraged for data as a product.

***Table 9-1.*** *Release Stages*

| RELEASE TYPE | RELEASE DESCRIPTION | PRODUCT LIFE CYCLE MOMENT AND POTENTIAL USE CASE |
| --- | --- | --- |
| 1  Pre-alpha | It's when the team designs the solution.<br><br>It englobes the analysis, design, development, and unit testing. | Idealization of the product<br><br>• It can be used to create the hypothesis design. |
| 2  Alpha | After the pre-alpha release, the software starts acting as the final product, with some functionality. | Idealization of the product<br><br>• It can be used for prototyping. |

*(continued)*

***Table 9-1.*** (*continued*)

| RELEASE TYPE | RELEASE DESCRIPTION | PRODUCT LIFE CYCLE MOMENT AND POTENTIAL USE CASE |
|---|---|---|
| 3  Beta | After the alpha release, it's closer to the final product.<br><br>It has more stable main functionalities but is still being tested with a selected target group of potential customers.<br><br>They are called beta testers and provide feedback on their experience with the solution. | Materialization of the product<br><br>• It can be used for prototyping validation. |
| 4  Release Candidate | After the beta release, the software applied all the beta feedback. Then, it's time for the team to apply for a complete test.<br><br>The goal is to make reliable software for a wider audience. | Materialization of the product<br><br>• It can be used for prototyping validation. |
| 5  General Release | After concluding the release candidate, it's put into production and available to the wide and target public.<br><br>Then, get feedback from the marketing and the whole target public. | Materialization of the product<br><br>• It can be used for the implementation and launch of the product. |

(*continued*)

***Table 9-1.*** (*continued*)

| RELEASE TYPE | RELEASE DESCRIPTION | PRODUCT LIFE CYCLE MOMENT AND POTENTIAL USE CASE |
|---|---|---|
| 6 Production Release | The final release leverages the improvements realized during the general release.<br><br>Customers use the software.<br><br>Then, it designs two types of releases.<br><br>• **Regular updates** maintain smooth operation.<br><br>• **Long-term** implements and supports new versions for users by ensuring consistency over time for customers. | Ongoing of the product<br><br>• It can be used for the next quadrant of the canvas, which is covered in the next chapter. |

# Defining the Release of the Product

An important topic that usually creates confusion or curiosity is related to the decision maker of releases. The answer is simple: the decision maker for the release moment is the owner of the product. In the data world: the data product manager.

For this reason, the definition of the strategy is completely related to your vision.

That's why it is important for you to understand the software release stages to map how it can be leveraged for your context if you desire it.

Besides, it'll also enable you to communicate with different teams or technical professionals that may use these nomenclatures. At the end of this stage, you should have clarity about the whole solution you'll consider and clearly document it in the Data Product Management Canvas. See Figure 9-1.

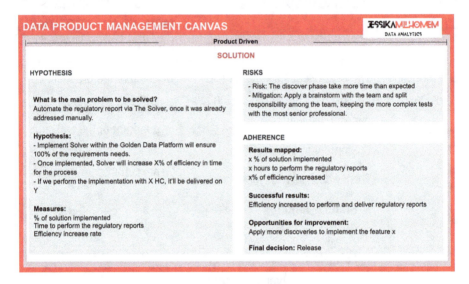

***Figure 9-1.*** *The Data Product Management Canvas: sample for idealizing and building the solution*

# 9.3.  Concluding the Project

You have built the product and made it available for all customers.

The definition of done must also include its full documentation and customer training, so it is critical to have it done as part of the process of making it available for the final customers.

Besides, everything starts with the project, and every time you implement improvements on the product, it'll cover the whole canvas again and require project management. Thus, you must ensure the following materials.

It's relevant for your customer and your sponsor, stakeholders, shareholders, and team.

# Full Documentation of the Product

Just as important as the product itself is the documentation of it. Because, after all, the goal is to empower your customer to solve their problems through the product you created.

## Agnostic to the Data Product

Regardless of the data product you created, it's crucial you create a complete document explaining the objective of the product. It should consist of the following.

- The objective of the product

- The reason for its existence

- The use cases it solves

- Reference to all the documentation mentioned in this section

## Data Product

If you created a data product, it's essential to have documentation explaining its architecture and functionalities. It should consider the following.

- The product's architecture: the minimum necessary for the user to use it

- The product's components: the components and how they work

- The product's features: what they are, what they solve, for what purposes

- A list of pre-requirements to use the solution

- Reference to the playbooks and other documentation mentioned in this section

## Data as a Product

If you created data as a product, it's essential to create documentation explaining its architecture and business contexts. It should consider the following.

- **The product's architecture**: The minimum necessary for the user to use it, such as the business domain definitions and layers of data architecture

- **The business domain context**: The business domains and subdomains related to the data as a product, and the business context

- **The product's features**: What they are, what they solve, for which purposes

- The list of pre-requirements to use the solution

- The reference to the playbooks and other documentation mentioned in this section

I recommend that you work with the Golden Data Platform to address data as a product need (see Chapter 5).

## Customer Playbooks

Playbooks are crucial, but I don't agree that they are optional, as many professionals point out, especially if you want the customer to have the self-service experience, which, honestly, is essential in my vision.

For this, the minimum guidelines must be available for the customers. It can be done through written, video, or mixed documentation. The important thing is to give clarity about the usage of it. For this, it should cover the following.

- Setting up the product

- How to get started using the product

- Each functionality of the product

## Training

It's important that you train your customers. Of course, the playbook is important, but from the customer's perspective, if it's necessary to go deeper, it's important to have other tools to learn about it.

The strategy of the availability of this material is up to you. If it's an internal solution, it'll possibly be free training, but if it's for external customers, it can be a monetary reward or free. Again, it's up to you to decide the strategy. It'll depend on the proposal you designed for your product.

## Optional: Hand over Assisted of Domains from the Golden Data Platform

If you are using the Golden Data Platform to implement the organization's domains as a whole, it may be necessary to work on handing it over to the appropriate team owner of the domain.

For this context, you must define work agreements with the involved professionals to ensure the process is performed as smoothly as possible.

## Results Presentation

Finally, it's fundamental that you create documentation and perform a presentation for the sponsor, and possibly also for the stakeholders, depending on your context, to present the results achieved with the project.

# 9.4. Measuring the Product

The product is built (at least the MVP of it) and in production. It's time to measure its results comparing with the process thus far! Yes, the whole process contains the measurement process. It's critical for the whole journey, not just at the beginning or at the end. Remember it's a cycle process.

## Adherence

Let's discuss the product's adherence measurement (see Figure 9-1).

The objective is to evaluate the results of the product after the launch of it.

The purpose of the results is similar to the previous one: evaluate what is working and keep it, possibly improving even more. Depending on the results, the strategy for the product will eventually be reevaluated and adapted.

Anyway, it's time to update the information necessary for the strategy. Even if it's just to update the actual results of the strategy designed. The important point is to keep the recurrence of the inspection of the full plan.

And who is responsible for that? The owner of the product: the data product manager.

For the adherence mapping, it's important to have clarity about the following.

- The method of validation

- The metrics that must be evaluated

- What are the thresholds and definitions of a successful diagnosis

- What are the thresholds and definitions of a failed diagnosis

- The evaluation period

Data tracked during the evaluation period is enriched with customers' data and behavior regarding the product's usage.

# Perspectives to Evaluate

The metrics designed for product strategy should enable you to evaluate the following perspectives.

## Problem Resolution

The first aspect to consider is related to resolving the business problem. Based on the data, you and/or your team should be able to see evidence that supports you in answering the following questions.

- **Purpose/value for the business**

    - Is the problem prioritized effectively solved with the project?

    - Is the resolution relevant to the business need as expected before?

    - What is the purpose of value generated so far? What is the evidence for that?

    - Are the results following the expectations planned, or do they have a different perspective?

    - Is that a good trend or a bad trend? Why?

- **Features and expectations**

    - Are the features performing as expected after the experiments?

    - Are there any issues that must be evaluated before the plan is designed?

    - Should we consider reprioritization?

## Customer's Perspective

The other aspect that is crucial to evaluate is the customer's perspective.

The objective is to get feedback on the usability and functionalities of the product, considering the product's proposal so far and the next steps. You can use some of the following questions.

- **Solution usability**

    - Was the behavior of the customer expected?

    - What is the positive feedback for the solution?

    - What is the negative feedback for the solution?

    - Does it solve completely the customers' needs from their perspective?

    - If not, what is not solved and why?

- **Mapping of any critical functionalities missing:**
  Based on the data analyzed, you may identify that some functionalities are missing and should be considered a priority by requesting a pivot of the plan, or they are improvements that enrich the designed solution. Anyway, this information is relevant. And you should ask yourself the following.

    - What functionalities are missing?

    - Are they hard blockers or improvements?

- **Validation of the customers' personas and segments:**
  Based on the feedback, it's also important to evaluate if we should keep the current plan, focus on some specific segment of base customers, or work with additional profiles.

- Should we keep the current plan?

- Should we prioritize the resolution of a specific segment of customers?

- Should we consider any additional personas or customer segments? If so, which? Why?

- What is the impact of the evaluation, and what should be done about it?

The goal of this phase is to have clarity on the result of the product design and be able to answer the primary question. Can we move forward on its strategy's evolution and next steps, or should we pivot the plan?

# 9.5. Summary

Let's recap the main takeaways of this chapter.

- **Understanding the issues and empathizing**: To design a product, the first step is to understand the business needs, which can be related to issues, challenges, or insights. It's fundamental to emphasize to the customers and get clarifications about the business context. Start the proposal project solution to design the product and address the business needs.

  We covered the approach to perform it through the Business Driven pillar of the Data Product Management Canvas.

- **Idealizing the data product**: As soon as the business needs are clarified and a proposal is made, explore and create the strategy for the product, considering not just the short term but also the middle and long term.

Then, with the short-term targets defined, it's time to go deeper into the tactical idealization of the product. And prototype the product.

You learned that approach by following the full Strategy quadrant and part of the Solution quadrant from the Data Product Management Canvas.

- **Materializing the data product**: Finally, it's time to build the product and make sure all the necessary work related to that to launch the product and have it available for the whole market. Test the prototype and put the product into production.

  You learned the approaches for it by concluding the Solution quadrant of the Data Product Management Canvas in this chapter.

## Summary

In this chapter we discussed the materialization of the product. It concluded coverage of the Solution quadrant by discussing tests, the strategy to launch a product, and how to measure its results.

With this knowledge, you can design and implement a product strategy.

The next chapter covers the last part of the Product Driven pillar in the Data Product Management Canvas. It discusses the ongoing product, which should ideally be considered before the product's launch and started as soon as it is in production.

# CHAPTER 10

# Ownership Process Recurrent Cycle: Ongoing

You can understand a problem by mapping the business needs through the data product management's Business Driven pillar. You learned how to idealize and materialize the product by first defining the strategy for the product and exploring and materializing the product design. The process for you to learn it was done through the coverage of the data product management's Product Driven pillar with the Strategic and Solution quadrants.

This chapter concludes the discussion of the Product Driven pillar by talking about the ongoing product.

Figure 10-1 features the post-release steps for data product management.

© Jessika Milhomem 2025
J. Milhomem, *Data Product Management in the AI Age*,
https://doi.org/10.1007/979-8-8688-1315-3_10

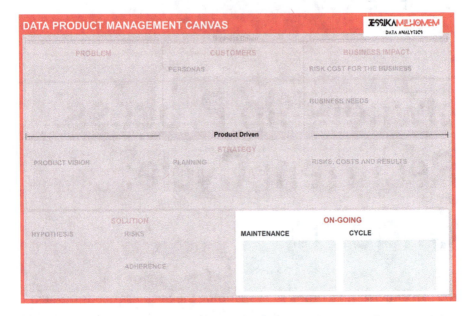

**Figure 10-1.** *The Data Product Management Canvas: the ongoing product*

Let's talk about the motivation and the work to be covered in each one of the quadrants. After reading this chapter, you can use the Data Product Management Canvas to design and visualize your ongoing strategy.

# 10.1. The Motivation

The ongoing process is a super important part of data product management that is commonly overlooked.

The following are the main reasons for its importance.

- Maintenance and enhancement or evolution of the product created

- Monitoring of new business needs

- Functional data ownership

- Healthy business processes related to or using
  the product

# Product Maintenance

It's fundamental to have professionals or mechanisms focused on maintaining the implemented product. The business must be observable due to the needs and issues addressed with the problems identified.

# Monitoring New Business Needs

The business is organic and is continuously evolving and changing. Consequently, new necessities or issues will appear, and it's important to be aware of these new potential scenarios to support the users. It's even more important for the business context the product is already addressing.

# Functional Data Ownership

Another reason for the importance of the ongoing is pragmatic data ownership. It's critical that data ownership is practiced by the teams that own data. It is even more important for data as a product. I'd say they are crucial.

# Functional and Healthy Business Processes

Finally, and also critical, is the insurance of healthy business processes related to or using the product created.

Similar to the previous context, it's decisive when discussing data as a product.

Imagine if there's some issue in some fundamental data for a specific domain. Now, imagine that there are a bunch of processes related to that, but no one cares about its resolution. Wouldn't it be chaotic and catastrophic?

That's the importance of clear ongoing definitions.

# 10.2. Maintenance

The Maintenance quadrant is where the following needs are defined.

- Work agreements

- Processes and tools

- Measures for observability

## Work Agreements

Work agreements are agreements you need to make with your customers and stakeholders. With customers, it can be just a thought process, while stakeholders and partners should be detailed and aligned.

For this, you can use the following list of questions to design, refine, or reinforce the resolutions.

- What should be considered maintenance, and what should be considered as the product cycle and evolution?

- Should someone outside of the team owner of the platform be involved in specific scenarios of issues or challenges? If so, who and why?

- Is it necessary to align any specific work agreements with peers? If so, what? Why? When?

- In the case of external customers, should any "agreement" be made? Is anything mandatory for legal or compliance reasons?

- What about stakeholders? And customer?

- What are the main work agreements to consider?

- How should be the internal organization of the team to support the maintenance requests and business needs?

- Who is responsible for maintaining it?

## Processes and Tools

Besides the work agreements, processes are relevant to ensure the relevance of the product availability. Processes and tools are relevant. You need to consider the following questions to define your strategy.

- What is the best process for customers to request improvements or share feedback?

- Should it be automated now, or can it be done later?

- If it's manual, how should the process be?

- Which tool should be used to consolidate all the demands from different fronts?

- Who will evaluate the requests? When?

- How should the processes be for the customer to open their requests?

- What is the process to evaluate and resolve the demands?

- Should it be necessary to define a service-level objective for the team to maintain it? Why? What should be the rules?

- Should the team define any service-level agreement? Why? What should be the rules?

# Measuring the Business as Usual (BAU)

Finally, it's time to measure and monitor the metrics related to daily product maintenance.

Ideally, it should be considered in the product strategy. If it's not, include it for a holistic vision of the product.

Table 8-3 presents business perspective metrics that you can map, including "internal perspective" and "innovative and learning" layers. Review the strategy to ensure a 360-degree perspective.

Besides, you should recurrently evaluate the results of the maintenance process and apply adaptations for improvements as soon as you see critical moments based on the thresholds you defined. At the end of this stage, you will keep the tracking of the metrics and evaluation and keep it clearly documented in the Data Product Management Canvas. See the Figure 10-2.

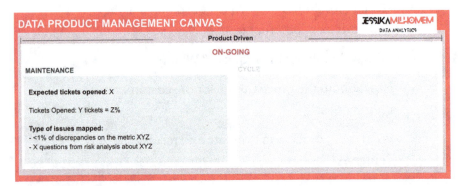

***Figure 10-2.***  *The Data Product Management Canvas: Maintenance Sample*

# 10.3.  Cycle

The Cycle quadrant is where the evaluations related to emerging needs are put in place.

This quadrant maps two questions.

- What should be done if new needs appear?

- Are there mapped problems and challenges that require us to restart the cycle?

## Emerging Needs

During the maintenance of the product and its processes, your team may identify requests for new needs.

Here, three characteristics must be mapped as macro perspectives for further evaluation by the product owner.

- What is the problem?

- Why is it a problem?

- Who is suffering from it?

## Next Steps

The resolution for the problem should be done by following the work agreements previously defined.

As soon as it is addressed, it returns to the data product manager the responsibility to keep the product's life cycle by the whole data product management process again. At the end of this stage, you should have clarity about the whole solution you'll consider and clearly document it in the Data Product Management Canvas. See the Figure 10-3.

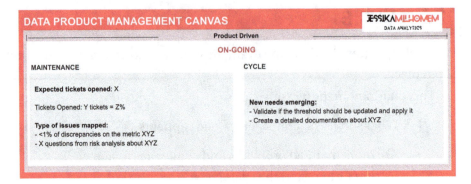

*Figure 10-3.* *Sample for the ongoing product*

## 10.4. Summary

Let's recap the main takeaways of this chapter.

- **Understanding the issues and empathizing**: To design a product, the first step is to understand the business needs, which can be related to issues, challenges, or insights. It's fundamental to emphasize to the customers and get clarifications about the business context. Also, start the proposal project solution to design the product and address the business needs.

  This book covered the approach to performing it through the Business Driven pillar of the Data Product Management Canvas.

- **Idealizing the data product**: As soon as the business needs are clarified and a proposal exists for that, it's time to explore and create the strategy for the product, considering not just the short term but also the middle and long term.

Then, with the short-term targets defined, go deeper into the tactical idealization of the product. And prototype the product.

You learned that approach by following the full Strategy quadrant and part of the Solution quadrant from the Data Product Management Canvas.

- **Materializing the data product**: Build the product and make sure all the necessary work related to it to launch it and have it available for the whole market. Test the prototype and put the product into production.

  You learned the approaches for it by concluding the Solution quadrant of the Data Product Management Canvas.

- **The ongoing data product**: Finally, after building and deploying the product, it's critical to define the maintenance and continuous cycle of the product. You learned the approaches for it by covering the Ongoing quadrant of the Data Product Management Canvas.

# Summary

This chapter consolidated what you learned about the Data Product Management Canvas with the "ongoing" quadrant. Finalizing that way is the complete practical approach to managing data products with the framework created and by aggregating knowledge on the Golden Data Platform.

Congratulations, you've completed the book! I hope you practice the material you learned in this book and, based on it, create and evolve your products by solving real problems! Also, enrich the data world's knowledge!

Be the difference! :)

I wish the best for you.

## 10.5.  Ready to go further?

If you're interested in mentorship, consulting, or upcoming courses on data product management and digital transformation, I invite you to connect. Fill out the form at ➤ https://jessikamilhomem.com/contact to be among the first to hear about new opportunities. I personally review every message.

# Index

## A

ACID, *see* Atomicity, consistency, isolation, durability (ACID)

Adherence, solution test
  Data Product Management
    Canvas, 334
  evaluation methods
    group interview, 335
    hands-on usability
      testing, 336
    online user interview, 336
    quantitative and qualitative
      information, 335
    user interview, 335
  evaluation perspectives
    customer's perspective,
      337, 338
    problem resolution, 337
  mapping, 334

AI/ML outputs, 5

Alert intelligence, 253

Analytics, 42
  experiences, 212
  innovations, 212
  issues and challenges, 211
  maintaining cycle, 212
  samples, 211

Apache Spark, 143

Apache Spark1, 122

Artificial general intelligence (AGI), 46, 47, 189

Artificial intelligence (AI), 189, 309, 312
  AGI/strong AI, 46, 47
  narrow AI, 45, 46
  superintelligent AI, 47

AI, *see* Artificial intelligence (AI)

Artificial neural networks (ANNs)
  backpropagation, 53
  feedforward, 52
  layers, 52
  nodes, 52

Atomicity, consistency, isolation, durability (ACID), 86

## B

Balanced scorecard, 313, 314

BI architecture, 100
  computing processing
    limitations
      ability to use real-time
      analytics to integrate with
      services with lower latency
      and large volume, 110

# G

GPSR Compliance
The European Union's (EU) General Product Safety Regulation (GPSR) is a set
of rules that requires consumer products to be safe and our obligations to
ensure this.

If you have any concerns about our products, you can contact us on

ProductSafety@springernature.com

In case Publisher is established outside the EU, the EU authorized
representative is:

Springer Nature Customer Service Center GmbH
Europaplatz 3
69115 Heidelberg, Germany

9 798868 813146